THE HUMAN SOUL

THE HUMAN SOUL

Explanation of the work
Quaestiones Disputatae de Anima
by Saint Thomas Aquinas

Miguel Grosso

First Edition. December 2024
Copyright © 2024 Miguel Alberto Grosso
ISBN: 9798302743329
grossomiguel2005@yahoo.com.ar
Independent Publication
All rights reserved

Original title: *El alma humana. Explicación de la obra
"Quaestiones Disputatae de Anima" de Santo Tomás de Aquino.*
(2024)
Author: Miguel Grosso

INDEX

INTRODUCTION 1
1. QUESTION 1: Whether the soul can be both a form and something in itself 5
2. QUESTION 2: Whether the human soul, so far as its act of existing is concerned, is separated from the body 16
3. QUESTION 3: Whether there is one possible intellect, or intellective soul, for all men 26
4. QUESTION 4: Whether it is necessary to admit the existence of an agent intellect 36
5. QUESTION 5: Whether there is one separately existing agent intellect for all men 44
6. QUESTION 6: Whether the soul is composed of matter and form 53
7. QUESTION 7: Whether the angel and the soul are of different species 63
8. QUESTION 8: Whether the rational soul must be united to a body like that which man possesses 74
9. QUESTION 9: Whether the soul is united to corporeal matter through a medium 84
10. QUESTION 10: Whether the soul exists in the entire body and in each of its parts 92
11. QUESTION 11: Whether the rational, sensitive, and vegetative souls in man are substantially one and the same 99
12. QUESTION 12: Whether the soul is its powers 108
13. QUESTION 13: Whether the powers of the soul are distinguished from one another by their objects 116
14. QUESTION 14: Whether the human soul is corruptible 126
15. QUESTION 15: Whether the soul, when separated from the body, is capable of understanding 134
16. QUESTION 16: Whether the soul, when united to the body, can understand separated substances 143
17. QUESTION 17: Whether the soul, when separated from the body, can understand the separated substances 149
18. QUESTION 18: Whether the soul, separated from the body, knows all natural things 155
19. QUESTION 19: Weter the sentient powers remain in the separated soul ..165
20. QUESTION 20: Whether the soul, separated from the body, knows individual beings..173
21. QUESTION 21: Whether the soul, separated from the body, can suffer the punishment of corporeal fire 183
BY WAY OF AN EPILOGUE..191
ENDNOTES 1

INTRODUCTION

1-What is a *Quaestio disputata*?

The *quaestio* was a fundamental teaching and academic debate method in medieval universities, especially within the realms of philosophy and theology. This structured format, which consisted of posing a question, analyzing different perspectives, and arriving at a well-reasoned conclusion, became the heart of scholastic education.

The roots of the *quaestio* lie in classical antiquity, particularly in the rhetorical and dialectical traditions. However, it was during the Middle Ages, especially in the 12th-century universities, that this method was solidified and became a key tool for knowledge transmission. The University of Paris, in particular, played a crucial role in formalizing the *quaestio*.

The *quaestio* developed through three main stages:

> **1-*Lectio***: The teacher selected a text of authority (such as the Bible or the works of Aristotle) and analyzed it in depth with the students
>
> **2-*Quaestio***: Based on the reading, a specific question requiring an argued response was formulated
>
> **3-*Disputatio***: A debate was organized in which students presented various arguments for and against the proposed thesis. The teacher, acting as an arbiter, guided the discussion and provided a definitive response

There are different types of *quaestiones*, each with its own characteristics:

> ***Quaestio disputata***: A formal debate in which two students defended opposing positions
>
> ***Quaestio quodlibetalis***: An open debate in which any member of the university community could pose a question
>
> ***Quaestio terminabilis***: A debate focused on a specific topic with a set duration

The teacher played a fundamental role in the *quaestio*. They were responsible for selecting the texts, formulating the questions, guiding the discussion, and offering the final response. Students actively participated in the debate, developing their analytical, argumentative, and synthesis skills.

The *quaestio* had a profound impact on the development of Western thought. Some of its major legacies include:

> **Promotion of critical thinking**: The *quaestio* encouraged students to analyze various perspectives, evaluate arguments, and form their own judgments
>
> **Development of communicative skills**: The debates fostered the ability to express ideas clearly and concisely and to respond to objections
>
> **Consolidation of knowledge**: The *quaestio* contributed to the systematization of knowledge and the construction of robust doctrinal frameworks

Although the *quaestio* as a formal teaching method is no longer used, its legacy lives on in contemporary education. Many of the principles of the *quaestio*, such as the importance of debate, critical analysis, and the search for truth, remain valid and are applied in various contexts, from classrooms to online discussion forums.

2-*Quaestiones Disputatae de Anima*

In the 1960s, James H. Robb (1918–1993), a professor of medieval philosophy at Marquette University, evaluated the sources and bibliography of *Quaestiones Disputatae de Anima* by St. Thomas Aquinas. He concluded that existing studies were insufficient for a thorough historical investigation. Determined to create a new critical edition, Robb traveled to Europe to consult original manuscripts and received support from renowned scholars like Étienne Gilson.

His work resulted in the first critical edition of the non-Parisian tradition of this work, published in Latin in 1968. Thanks to this edition, the current Spanish translation, based on Robb's text, became possible.

The first printed edition of *Quaestiones Disputatae de Anima* dates back to Venice in 1472. Over time, various editions emerged, notably the *Piana* edition of Rome (1570–1571), promoted by Pope Pius V during the Council of Trent. While this edition aimed to preserve the authenticity of Aquinas' texts and prevent doctrinal alterations, it was not a critical edition but rather a compilation based on existing texts. Its influence was significant, as it served as the foundation for subsequent editions in Antwerp, Paris, and Parma, which, however, also lacked exhaustive critical review.

Only in the Paris edition by the Vivès publishing house (1871–1882) were traditional manuscripts used, particularly for the *Summa Theologiae* and *Summa contra Gentiles*, although the *Quaestiones Disputatae* retained the unaltered text of the *Piana*.

Robb's edition is thus crucial, as it is the first critical edition distinct from the *Piana*. Regarding authenticity, Robb himself states: no one has questioned the authenticity of *Quaestiones Disputatae de Anima*. Since the late 13th and early 14th centuries, it has appeared in the catalogs of the University of Paris, as well as in those of Ptolemy of Lucca, Bernard Gui, Bartholomew of Capua, Nicholas Trivet, John of Cologne, and William of Tocco.

The work *Quaestio Disputata de Anima*, or *Single Question on the Soul*, has traditionally been named according to the Roman edition of 1570. However, historical research shows that in manuscripts and early editions, it was considered a series of 21 Questions, not a single Question divided into articles. This fact supports the title *Quaestiones de Anima*, which is more faithful to the manuscript tradition.

The exact date when Aquinas disputed these Questions is unknown. However, evidence suggests they were debated in Paris in 1269. These public disputes, conducted alongside advanced students, were formalized by the master after their celebration. All of Aquinas' *Quaestiones Disputatae* were drafted by him personally, not as third-party reports, which adds authenticity to the texts.[1]

1. QUESTION 1: Whether the soul can be both a form and something in itself

> Saint Thomas presents the arguments of various authors, according to which it seems that the human soul cannot be *hoc aliquid* ("something in itself") and at the same time the form (principle of organization) of the body

1- If the human soul is *hoc aliquid* or "something in itself" (i.e., an independent substance), it possesses a complete being of its own. Following this logic, anything added to a complete being is joined to it accidentally, like the quality of whiteness in a person or clothing as a possession. Therefore, if the soul is something independent and complete, the body would be joined to it accidentally, as something external, and consequently, the soul could not be the substantial form of the body. This would imply that the soul would not truly integrate as the constitutive principle of the human body, but rather as an entity separate and external to it.

2- If the soul is something in itself *(hoc aliquid)*, it must be something individual, as no universal can be something in itself. This individuation of the soul must come either from something else or from itself. If the individuation comes from something else and the soul is the form of the body, then this individuation would come from the body (as forms are particularized or individuated through their matter). Consequently, when separated from the body, the individuation of the soul would vanish. This implies that the soul cannot exist as a separate and individual substance.

If the soul individuates by itself, two possibilities arise: either it is a simple form (without composition of matter and form) or it is something composed of matter and form.

A. If it were a simple form, individuated souls could only differ from one another by their form, which would imply a difference in species between the souls of different individuals. This would lead to the conclusion that humans would differ in species, which contradicts the notion of shared humanity.

B. If the soul is composed of matter and form, then it could not be the form of the body, as matter cannot be the form of something. This also contradicts the idea that the soul could simultaneously be an independent entity *(hoc aliquid)* and the form of the body.

3- If the soul is an "individual in itself," it must belong to a specific species and genus. However, if the soul has its own species and genus, it would be impossible for this soul (with its own specific nature) to unite with the body to form a new species (the human being) without changing the nature of the union.

If the soul already possesses its own species and genus, it could not unite with the body as a form, since a pre-existing form or species does not accept an "addition" of another without losing its original identity (as Aristotle states, forms function like numbers; if something is added or subtracted, their essence changes). Thus, the soul could not act as the form of the body because it would already be a separate species, making union with the body to constitute a new species impossible.

4- God, in His goodness, created the universe with a hierarchy of beings, each occupying a specific level. If the human soul could subsist on its own as "something in itself" *(hoc aliquid)*, it would occupy a level in the hierarchy of beings. However, forms by themselves, without matter, do not constitute levels of separate beings. Therefore, if the soul is "something in itself," it could not be the form of matter, such as the human body.

5- If the soul is "something in itself" and subsists by itself, it must be incorruptible, as it is not composed of contraries (opposites), which is characteristic of corruptible beings. However, the human body is corruptible, and in scholastic doctrine, a form must be proportionate to its matter. Thus, if the soul is incorruptible, it could not be the form of the corruptible human body, as there would be no proportionality between them.

6- Only God is pure act, that is, full existence without potentiality. If the soul were "something in itself," independent and subsistent, it would possess a combination of act and potency. But since a form cannot be in potency, the fact that the soul has some degree of potentiality would disqualify it from being the form of the body.

7- The soul unites with the body for some utility, whether essential or accidental. Essential union is not necessary, as the soul can exist without the body. An accidental union also does not seem likely, since the main accidental benefit, knowledge acquired through the senses, does not hold; some believe that the souls of children who die before birth possess perfect knowledge without having experienced the senses. This suggests that, if the soul is "something in itself," it has no reason to unite with the body as a form.

8- According to Aristotle, substance is divided into three opposing parts: form, matter, and *hoc aliquid* (a "something in itself"). Since form and *hoc aliquid* are opposites, they cannot coincide in the same being. It follows that the human soul cannot be simultaneously form and *hoc aliquid*.

9- Here it is argued that an entity that is "something in itself" must exist independently, while the nature of a form is to exist in something else (i.e., to be part of something else). As subsisting by itself and existing in something else are opposite qualities, if the soul is "something in itself," then it cannot be the form of the body.

10- If the soul subsists after the death of the body and, in that state, loses its nature as a form, then the characteristic of "form" is accidental to the soul. However, the soul only unites with the body as a form, and if this union is accidental, the human being would not be an "essential being" but an "accidental being." This is inconsistent with the view of the human person as a substantial unity of body and soul, which is deemed unacceptable in Thomistic philosophy.

11- If the human soul is *hoc aliquid* (an independent entity existing on its own), it should have its own operation, as every entity that exists by itself performs some particular activity. However, it is argued that the intellectual operation *(intelligere)*, which seems proper to the soul, does not belong exclusively to the soul but to the human being as a composite of body and soul. This leads to the conclusion that the soul is not *hoc aliquid* in the sense of an independent entity.

12- If the human soul is the form of the body, it should depend on the body to some extent, as form and matter are interdependent. Thus, what depends on another is not *hoc aliquid* (a completely independent entity), so if the soul depends on the body, it cannot be considered *hoc aliquid*.

13- Since the soul and the body belong to different categories (the soul belongs to the category of incorporeal substances, while the body is a corporeal substance), they cannot share the same "being." Therefore, the soul could not be the form of the body if its "being" is not shared with the body.

14- The being of the body is corruptible and composed of quantitative parts, while the being of the soul is incorruptible and simple. Thus, the soul and the body do not share the same "being," which suggests that the soul cannot simply be the form of the body.

15- Even if it is argued that the human body obtains its "being" through the soul, a response can be made by citing Aristotle, who holds that the soul is the act of an organic physical body. Consequently, the body, as an organic subject, must already be a body constituted within the corporeal category by some form. Thus, the human body possesses its own "being" separate from the soul.

16- Essential principles like matter and form are oriented toward "being." When something can be completed with a single principle, two are not required. If the soul has its own "being" as hoc aliquid (a distinct entity), the body would not naturally unite with it except as matter that acts as a support for the form.

17- The "being" (= act of existing= *actus essendi*) acts in the substance of the soul as its act, and should be supreme within it. Since the lower does not reach the higher at its supreme part but at its lower level, the body (as inferior to the soul) could not participate in the "being" that is supreme in the soul.

18- Things that share the same "being" also share a single operation. If the esse (= act of existing= *actus essendi*) of the human soul united with the body were common with the body, then its operation, which is intelligere (understanding), would also have to be common to the body, which is impossible, as Aristotle demonstrates. Therefore, the soul and the body cannot have the same "being," indicating that the soul cannot simply be the form of the body and hoc aliquid.

> Next, Saint Thomas presents two arguments from authority, according to which the soul can be a form and at the same time something in itself

1- Each thing obtains its species through its proper form. In the case of the human being, its identity and specific essence reside in its rationality, that is, its capacity to reason. Therefore, it is concluded that the rational soul is the specific form of the human being because it is rationality that essentially defines what man is. Thus, the rational soul is presented as the form that makes man what he is, giving him his specific identity.

It is further noted that the rational soul is *hoc aliquid* (a subsistent entity) because it operates independently. Specifically, the intellectual capacity for understanding (in particular, the abstraction of the essence of beings) does not depend on a bodily organ. In *De Anima*, Aristotle states that the intellectual operation is independent of the body. This means that the human soul possesses its own existence and is capable of operating without the body. Consequently, it is asserted that the human soul can be both *hoc aliquid* and the form of the body, as it is independent in its intellectual operations, but at the same time defines the human essence.

2- Finally, it is argued that the ultimate perfection of the human soul lies in its ability to know the truth, which occurs through the intellect. To achieve this perfection in the knowledge of truth, the soul needs to be united to the body, as it depends on the *phantasmata* (sensible images) generated by the bodily senses. The *phantasmata* are images or representations of perceived things that are essential to the process of intellectual knowledge. Since these images do not exist without the body, the soul must unite with the body as its form to be able to perform this operation.

> Next, Saint Thomas provides his own answer to the Question raised about the nature of the soul and its relationship with the body

In this sense, he presents the human soul as something unique in its kind, being both the form of the body and a subsistent entity *(hoc aliquid)*, capable of operating independently of matter. This argument combines Aristotelian principles with Christian ideas and is developed as follows:

1- The concept of "hoc aliquid" and its relation to the human soul. Saint Thomas begins by clarifying the concept of *hoc aliquid*, which refers to a concrete individual within the category of substances. Following Aristotle, he explains that primary substances *(primae substantiae)* are undoubtedly *hoc aliquid*, while secondary substances *(secundae substantiae)*, though they appear to have a similar entity, actually express a quality or essence common to several individuals.

In the case of the human soul, it can not only subsist by itself but is also something complete and specific in the category of substances. Although it unites with the body, the soul retains its independence and unique characteristics, distinguishing it from other material forms that do not have self-subsistence, such as body parts (hand or foot).

2- Rejection of alternative theories on the nature of the soul. The Angelic Doctor opposes the ideas of Empedocles and Galen, who viewed the soul as a harmony or combination of bodily qualities. For Saint Thomas, these ideas do not adequately explain the nature of the soul's operations, such as growth and nutrition in plants (vegetative soul) or sensory perception in animals (sensitive soul), as these require a principle higher than mere material qualities.

Moreover, these theories are even less satisfactory in explaining the rational soul, whose intellectual activity, including understanding and the abstraction of universal concepts, transcends material limitations and requires independence from the body.

3- The rational soul as subsistent and united to the body as its form. Saint Thomas maintains that the rational soul not only subsists by itself but also performs operations, such as understanding (the abstraction of essences), that are completely independent of the body. This is because understanding does not need a bodily organ to function, which demonstrates that the rational soul has a mode of existence independent of the body.

Following Aristotle and Plato, he affirms that the intellect is an incorruptible substance. Plato regarded the soul as immortal and self-moving. However, St. Thomas does not fully agree with him, as Plato understands the human being as the soul that "inhabits" a body, similar to a sailor in a ship.

4- The relationship between the soul and body in human beings. St. Thomas explains that the soul is what gives life to the body, and this union is not merely accidental. The soul is the form of the human body, imparting its being and species to the whole body and its parts. When the soul departs, the body parts lose their original names and functions, as they no longer fulfill their vital role. This demonstrates that the union between the soul and the body is essential.

Death, which is the separation of soul and body, implies substantial corruption, reinforcing the idea that the soul is the substantial form of the body, not an accidental form.

5- The human soul, midway between bodily and separate substances. He concludes that the human soul, insofar as it is united with the body, has a nature situated between purely material and purely spiritual substances. This union allows the human soul to perform an operation that transcends matter and, at the same time, to achieve a unique perfection in the knowledge of both the material and the universal.

The existence of the human soul, therefore, is elevated above the body, although it needs the body for its complete perfection and full operation in the human species. This dual nature places the human soul at an intermediate level between purely material and purely spiritual substances.

> Following this, St. Thomas responds to each of the eighteen initial arguments suggesting that the human soul cannot be both a form and something in itself

1- First objection. St. Thomas explains that, although the soul has "complete being," this does not mean that the body is united to it accidentally. The very "being" of the soul is shared with the body, creating unity in the being of the entire composite. Furthermore, although the soul can subsist on its own, it does not have a "complete species" without the body, which is necessary for its perfection.

2- Second objection. Every entity has being (act of existing=*esse*) and individuation simultaneously. Universals exist in reality only when individualized. Thus, the existence of the soul is given by God as the active cause and is in the body as in its matter, without depending on it for its endurance after the body dies.

3- Third objection. The human soul is not a *hoc aliquid* like a substance with a complete species, but rather as a part that composes a complete species, as previously explained. Therefore, the argument does not hold.

4- Fourth objection. Although the human soul can subsist on its own, it does not possess a complete species, so separated souls would not form a distinct "degree of being" like other complete beings.

5- Fifth objection. The human body is the matter proportioned to the soul, relating to it as potential to act. It is unnecessary for them to be equal in virtue of being—meaning they need not share an identical nature or level of perfection—since the soul is not a form fully contained by matter. This distinction is evidenced by the fact that some operations of the soul, such as intellectual understanding, surpass the limits of matter. According to faith, the body was created incorruptible, but due to sin, it was subjected to death, from which it will be freed in the resurrection.

6- Sixth objection. The human soul, being subsistent, is composed of potency and act. Its essence *(essentia)* is not its being *(esse=actus essendi)*, but it relates to it as potency to act. This does not prevent the soul from being the form of the body, as in other cases something that is act in one respect can be potency in another.

7- Seventh objection. The soul unites with the body for substantial perfection (to complete the human species) as well as for accidental perfection, as the acquisition of intellectual knowledge comes from the senses. Although the souls of children or the deceased have another mode of knowledge, this is due more to the separation than to human essence.

8- Eighth objection. It is not necessary that whatever is *hoc aliquid* be composed of matter and form, only that it can subsist by itself. Although the composite is *hoc aliquid*, this does not prevent other things from also being *hoc aliquid*.

9- Ninth objection. Something that exists in another as an accident in a subject loses the quality of *hoc aliquid*. However, something that is in another as a part does not necessarily lose that quality, like the soul in man.

10- Tenth objection. When the body perishes, the soul does not lose its nature as a form, even though it no longer actualizes matter in act, as it continues to be a form in potency.

11- Eleventh objection. Understanding (the abstraction of essences) is a proper operation of the soul as a principle that does not depend on the body. However, the body participates in understanding from the object's point of view, as images *(phantasmata)*, the object of the intellect, require bodily organs.

12- Twelfth objection. The soul depends on the body insofar as it needs the body to complete its species, but not to such an extent that it cannot exist without it.

13- Thirteenth objection. For the soul to be the form of the body, the "being" of the soul and body must be common, which is the "being" of the composite. This is not impeded by the difference in genera between soul and body, as both belong to a genus only as parts of the composite.

14- Fourteenth objection. What properly corrupts is neither the form nor the matter, but the composite. The body is said to be corruptible in that it loses the "being" it shared with the soul, which subsists by itself.

15- Fifteenth objection. In the definitions of forms, sometimes the subject in potency is used, such as when saying that motion is the act of what is in potency. Similarly, the soul is the act of the organic body, making it an organized body.

16- Sixteenth objection. The essential principles of a species are ordered not only to "being" in general but to the being of that particular species.

Although the soul can exist by itself, it cannot fulfill its species without the body.

17- Seventeenth objection. Although being is the most perfect form, it is also the most communicable. The body participates in the being of the soul, although not in as noble a way.

18- Eighteenth objection. Although the "being" of the soul is, in some way, shared with the body, the body does not partake in all the nobility and virtue of the soul's "being." Thus, there are operations of the soul in which the body does not participate.

2. QUESTION 2: Whether the human soul, so far as its act of existing is concerned, is separated from the body

> Saint Thomas presents the arguments of various authors, according to which it seems that the human soul is separated from the body in terms of its being (act of being = act of existing = *actus essendi*).

1- It is mentioned that in *De Anima* III, *The Philosopher* (Aristotle) states that no sentient power exists without a body. But the intellect is separate and the intellect is the human soul. Since the intellect is identified with the human soul, it is concluded that the human soul is also separated from the body in its act of existing.

2- This argument states that the soul is the act of the organic physical body and that the body is its instrument. If the intellect were united to the body in its act of being as a form, then the body would need to serve as its instrument, which Aristotle considers impossible. This implies that the intellect cannot be united to the body in the same way that a form unites with matter.

3- The union of form with matter is more intense than the union of a power with its organ. Since the intellect is simple and cannot be concretely united to the body as a power unites with its organ, it is concluded that it is even less likely that the intellect can be united to the body in the manner that form unites with matter.

4- This argument addresses the relationship between the intellect, understood as the intellective power, and the body, starting from the premise that the intellect does not possess a physical organ. It is established that, unlike other powers of the soul that rely on an organic body for their function, the intellect operates without the need for one. It is suggested that the essence of the intellective soul could unite with the body as a form, but this notion conflicts with the idea that the intellect cannot be an act of the body. Additionally, it supports the principle that "the effect is not simpler than its cause." This means that if a power, such as the intellect, is an effect of the soul's essence, it cannot be simpler than the essence of that soul, implying that the essence of the soul has a complexity not found in the individual powers that depend on it. Therefore, since the intellect cannot be the act of the body, it is concluded that the intellective soul also cannot unite with the body as a form. Thus, the intellect, as a non-individualized form, cannot have a union relationship with the body similar to that of a form with its matter. This argument reinforces the idea that the intellect has a nature essentially separate and distinct from the body, emphasizing its independence and operation outside corporeality.

5- It is argued that any form united with matter is individualized by it. If the intellectual soul were united with the body as a form, it would need to be individual. This would imply that the forms received by the soul would be individualized, which is problematic because it would negate the soul's ability to know the universal.

6- This argument states that the universal form cannot be intelligible from something external to the soul, given that all forms in external objects are individualized. If the forms of the intellect are universal, they must arise from the intellectual soul, implying that the soul is not an individualized form and, therefore, is not united to the body in its act of existing.

7- It is argued that intelligible forms, in relation to the soul, are individualized, but in their similarity to things, they are universal. However, since form is the principle of operation, the operation that follows would be individual and not universal, which is contradictory.

8- This refers to an assertion by Aristotle regarding the hierarchy among the different functions of the soul. Just as a triangle is in a square only potentially, the nutritive and sensitive parts are only potentially within the intellective part. Thus, since the nutritive and sensitive parts are not in act within the intellective part, it is concluded that the intellective part is not united to the body.

9- It is mentioned that one cannot simultaneously consider an animal and a man; one is first an animal and then a man. This implies that what constitutes the animal (the sensitive part) and what constitutes man (the intellective part) are not the same, reinforcing the idea that the sensitive and intellective parts are not united in a single substance.

10- It is established that form must be in the same genus as the matter it joins. Since the intellect does not belong to the genus of bodies, it is concluded that the intellect cannot be a form united to the body as matter is.

11- It is argued that from two substances that exist in act, a single one cannot be formed. Both the body and the intellect are substances existing in act. Therefore, the intellect cannot be united to the body to form a single thing from them.

12- Here it is argued that every form united with matter is realized through the movement and mutation of matter. However, the intellective soul is not realized from the potency of matter but rather comes from an external source, as stated by Aristotle in *De Anima* XVI. This implies that the intellective soul is not a form united with matter.

13- This argument states that every entity acts according to what it is. The intellective soul can act independently of the body, specifically in understanding. Therefore, it is not united with the body in its act of being or existing.

14- It is asserted that what is minimally inconceivable is impossible for God. It is considered inconceivable for an innocent soul to be enclosed in a body, similar to a prison. Therefore, it would be impossible for God to unite the intellective soul to the body.

15- Here it is mentioned that no wise artisan hinders his own work. However, the body is the greatest impediment to the intellective soul perceiving the truth, in which its perfection lies. This relates to the idea that the corruptible body weighs upon the soul. Therefore, God did not unite the intellective soul to the body.

16- It is argued that things united to one another have a mutual affinity. However, the intellective soul and the body are contrary, as the flesh desires the opposite of the spirit and vice versa. Therefore, the intellective soul is not united to the body.

17- This argument asserts that the intellect is in potentiality with respect to all intelligible forms, without having any in actuality; similarly to prime matter, which is in potentiality to all sensible forms. This implies that, just as there is a single prime matter for all things, the intellect is also one and, therefore, not united to the body, which individualizes it.

18-. It refers to Aristotle's statement in *De Anima* III, where he argues that if the possible intellect had a bodily organ, it would have some determined nature of the sensible natures and, therefore, could not receive and know all sensible forms. If the intellect were united to the body as a form, it would have to have a determined sensible nature and, thus, could not be receptive and cognizant of all sensible forms, which is impossible.

19- This argument establishes that every form united to matter is present in the received matter. What is received from something is in it according to the mode of the receiver. Therefore, every form united to matter is in it according to the mode of the matter. However, the mode of sensible and bodily matter does not allow receiving something intelligibly. Since the intellect has an intelligible being, it is not a form united to bodily matter.

20- It is argued that if the soul is united to bodily matter, it must be received in it. Everything that is received by something is received in its matter. Therefore, if the soul is united to matter, everything received in the soul is received in the matter. But the forms of the intellect cannot be received by prime matter; rather, they become intelligible through the abstraction from matter. Therefore, the soul united to bodily matter will not be receptive to intelligible forms, which implies that the intellect, which is receptive to these forms, is not united to bodily matter.

> Next, St. Thomas presents two arguments from authority that demonstrate the human soul is not separated from the body in terms of being (or act of being, or act of existing, or *actus essendi*)

The first argument refers to a statement by Aristotle in his work *De Anima*, where he suggests that one should not question whether the soul and the body are a single entity, just as one should not question the relationship between wax and its figure. Just as the figure cannot exist separately from the wax that gives it form, it is concluded that the soul cannot be separated from the body in terms of its existence. Since the intellect is considered a part of the soul, according to Aristotle, it is deduced that the intellect also cannot exist separately from the body.

The second argument reinforces this idea by asserting that no form can exist separate from its matter in terms of its act of being or existing. It is established that the intellectual soul is the form of the body, which implies that its existence is intrinsically linked to the matter of the body. Therefore, since the intellectual soul cannot exist without the matter it gives form to, it is concluded that it cannot be considered separate from the body in its being.

> Next, St. Thomas offers his own response to the Question raised

1- Consideration of the principle in potentiality and actuality. St. Thomas begins by affirming that where there is something that can be in potentiality and actuality (that is, that can be or not be at a given moment), there must be a principle that allows that thing to be in potentiality. He uses the example of a human being, who can be sensing in actuality or in potentiality. For the human being to sense, there must be a sensible principle in them that allows them to be in potentiality with respect to the sensibles. If they were always sensing in actuality, a sensible form would always have to be present.

2- Relation of the intellect to potentiality. In the same way, the human being can understand in actuality or be in potentiality to understand. Therefore, it is necessary to consider an intellectual principle in the human being that is in potentiality with respect to the things they can understand. This principle is what Aristotle calls the "possible intellect." This intellect must be in potentiality to receive the intelligible forms, just as the eye is in potentiality to receive all colors.

3- Nature of the possible intellect. St. Thomas concludes that the possible intellect must be "denuded" (stripped) of all sensible forms, which implies that it does not have a specific bodily organ. If it had an organ, it would be determined to a sensible nature, just as vision is determined by the eye. Therefore, the possible intellect cannot be similar to the sensible powers and should not be confused with them.

4- Refutation of other positions. Some ancient philosophers argued that the intellect was not different from the sensible powers, and others thought that the intellect was a form or virtue mixed with the body. However, these positions are refuted, for if the possible intellect were a substance separated from the body, it would be impossible for the person to understand through it. **The action of the intellect is completely different from the action of an external principle**. Therefore, if the intellect were separate, it could not act in the human being.

5- Clarification of the relationship between the intellect and images. St. Thomas mentions the idea that, although Averroes claimed that the possible intellect is a separate substance, he sought to connect the possible intellect with the images *(phantasmata)* that the human being generates from their sensory experiences. However, while this relationship suggests a connection, it is not sufficient to establish that the intellect is capable of understanding effectively.

6- Distinction between cognitive potentiality and species. The fact that the cognizable species are present does not mean that one can understand. Understanding depends on having a cognitive potentiality, which in this case is the possible intellect, which is not mixed with the body. Therefore, even though the images (phantasms) are accessible, they do not imply that the intellect comprehends.

7- Nature of the separate substance. Finally, St. Thomas holds that separate substances, being perfect, do not require material actions. The possible intellect, which is in potentiality to the species of sensible things and depends on human activity, cannot be one of those separate substances.

In conclusion, Saint Thomas establishes that the human soul is a form united to the body, but not entirely absorbed by it. Humanity has a capacity for understanding (potency) that is related to the possible intellect. This means that the intellect, although it does not depend on a physical organ, is manifested through the essence of the human soul, which is the form of the human being.

The soul is not separated from the body in terms of its essence. By defining the nature of the possible intellect and its relationship to the body, he refutes the idea that the intellect could exist as a separate entity and affirms that, in fact, the human intellect depends on the body and sensory experiences for its understanding. In this way, he defends the unity of the soul and body in the human being, contradicting views that consider the intellect to be an independent substance.

> Saint Thomas then responds to each of the twenty arguments initially presented, according to which the human soul is separated from the body in terms of being or existing

1- Saint Thomas explains that the intellect is said to be separate because it remains even when the body is corrupted; that is, the intellect can exist without the body, unlike the sentient powers that depend on the body to operate. The intellect does not require a bodily organ for its functions, which distinguishes it from the senses.

2- The human soul is considered the act of the organic body, for the body is its organ. However, the soul does not need the body to be its organ for the exercise of all its faculties, because the soul itself exceeds the proportion of the body. This means that, although the soul acts in the body, its nature is higher.

3- An organ is the principle of the operation of a power. If the intellect were united to an organ, its operation would depend on that organ. But since the human intellect is a virtue of the soul, it is not limited to a sensitive nature and, therefore, can operate without depending on a material organ.

4- The intellect relates to the soul in its ability to rise above the bodily matter. It is not the act of a specific organ but rather an essential part of the soul. Thus, although the intellect does not totally depend on the body, it is in harmony with the essence of the soul.

5- Although the human soul is an individualized form and has powers like the intellect, this does not prevent these powers from acting in an immaterial way. Separate forms can be individualized, and the intellect can comprehend the immaterial and universal, despite its individuation.

6- The intellect gives to the understood forms the capacity to be universal by abstracting them from the material principles that individualize them. Therefore, the intellect does not need to be universal in itself, but rather immaterial.

7- The species of an operation is derived from the form that is its principle. The effectiveness of the operation depends on how the subject is perfected. Thus, understanding the universal is part of the intellectual operation, and the way it is carried out determines its perfection.

8- The analogy between the parts of the soul and the figures shows that, just as a more complex figure includes what a simpler figure has, the sensitive soul contains what the nutritive soul has. This does not mean they are different in essence, but that there is a hierarchical inclusion.

9- The distinction between the concept of animal and the concept of man does not imply that there are different principles in each being. In animals, imperfect operations are evident before the more perfect ones, similar to how forms are generated.

10- The form does not belong to a specific genus; the intellectual soul is the form of man, and although united to the body, both are considered within the genus animal and the human species.

11- From two complete and perfect substances, one cannot generate a single one. However, the soul and body are parts of human nature, which allows them to form a unity.

12- Although the soul is a form united to the body, it exceeds the proportion of all bodily matter. Therefore, it cannot be entirely actualized by material movements or changes like other forms.

13- The soul has an operation that it does not share with the body in terms of its superior nature. However, this does not mean it is completely separated from the body.

14- This objection comes from Origen's position, who claimed that souls were created without bodies and later united to them. This is incorrect, as union with the body does not harm the soul, but perfects it.

15- The natural mode of knowledge of the soul involves perceiving intelligible truths through the senses. However, the corruption of the body affects this ability due to original sin.

16- The struggle between the carnal and spiritual indicates the connection of the soul with the body. The parts of the soul that are united to the body tend toward pleasure for the flesh, which can clash with the desires of the spirit.

17- The possible intellect does not have intelligible forms in act, but in potency. Therefore, it is incorrect to say that it is one in all, but rather that it is one in relation to all intelligible forms.

18- If the possible intellect had a bodily organ, that organ would be the principle of understanding, but this is false, as the intellect is not determined to a specific sensitive nature.

19- Although the soul is united to the body in a bodily way, its part that surpasses the body's capacity has an intellectual nature. Therefore, the forms received in it are intelligible and not material.

20- The response to this argument reaffirms that the soul, although united to the body, possesses an intellectual nature that differentiates it and allows it to understand immaterial reality.

Through these responses, Saint Thomas defends the nature of the human being, where the soul and body interact and complement each other, but also points out the superiority and independence of the human intellect in its capacity for knowledge and understanding.

3. QUESTION 3: Whether there is one possible intellect, or intellective soul, for all men

> St. Thomas presents the arguments by which it seems that the possible understanding (intellect) or the human intellectual soul is one in all human beings

1- Perfection must be proportional to the object it perfects. Truth is the perfection of the intellect, and since this truth is one and shared by all, some suggest that the possible intellect should be one in all human beings.

2- Saint Thomas quotes Saint Augustine, who expresses doubts about whether there is a single soul in all or multiple souls in many. He notes the difficulty of having the same soul simultaneously in a state of bliss and suffering in different individuals. He also finds it absurd to claim that there are multiple souls in multiple people. *Augustine says in the book De quantitate animae: "Concerning the number of souls, I do not know what to answer you."*

3- Every distinction between two things depends on having a determined nature. Since the possible intellect is potential in relation to all forms and lacks an actual form, it should not be limited or distinguished, and therefore, it could not be multiplied in several individuals.

4- Here it is established that the possible intellect is completely separate from what it understands, even from itself, and therefore lacks a basis to be multiple in different people.

5- Everything that is distinguished and multiplied must share something in common, like a genus. However, the possible intellect shares nothing in common with anything else, which implies that it cannot be distinguished or multiplied into different individuals.

6- Maimonides argues that beings separated from matter are only multiplied by cause and effect. Since one person's intellect is not the cause of another's intellect, and the possible intellect is a separate reality, it should not be multiple.

7- Aristotle teaches that the intellect is the same as that which is understood. Since the object of understanding is the same for all, it seems that the possible intellect is one and the same in all human beings.

8- The object of intellection is the universal, which is one in many. Since this unity does not come from the reality of individuals but from intellectual activity, it is deduced that the intellect must be one in all.

9- The argument of the common place of the soul. The text begins with a quote from *The Philosopher* in his work *De Anima*, where he asserts that the soul is the "place of the species" (understanding "species" as forms or universal concepts). The notion of "place" suggests something that can contain many things, so if the soul is the place of the species, it should be one and common to all human beings. This raises an objection to the idea that each person has an individual soul, since the concept of "place" applies to something that can house multiple entities without being multiplied for each one.

10- Some object to the intellect being the "place of the species." This term means that the intellect has the ability to contain the forms or universal concepts of the things we perceive and imagine. The objection suggests that if the intellect is considered the "place of the species" just because it "contains" those forms, then the same term should also apply to the senses. This is because the senses also "contain" or capture the forms of sensible objects, such as colors, sounds, etc., when we interact with them.

A response to this objection points out that Aristotle restricts this capacity of containment exclusively to the intellect and not to the senses. According to Aristotle, the intellect can contain the universal concepts of things, while the senses only perceive the particular and concrete. Thus, the intellect is the "place of forms" because it has the ability to abstract and understand universals, something the senses, which are limited to capturing the individual and particular, cannot do.

11- Since the intellect acts everywhere, knowing realities that exist anywhere, it seems to be present everywhere and therefore unique in all.

12- Everything that is particular requires specific matter to be individualized. Since the possible intellect is not tied to any matter, it is not defined as particular, suggesting that it is one in all.

13- Against this, it is proposed that the intellect is limited by the human body in which it resides. This assertion is refuted by claiming that, being external to the essence of the possible intellect, the body cannot be the principle of individuation or multiplication of the intellect.

14- According to Aristotle, if there were multiple worlds, there would be multiple first movers, which would be material, something impossible. By analogy, if there were multiple possible intellects, the intellect would be material, which is inadmissible.

15- If the intellects were multiple, they would remain after death, and thus differ in species. Since this would imply that humans have different species, which is clearly false, it is concluded that the intellect cannot be multiple.

16- What is separated from the material cannot be multiplied according to bodies. Since the intellect is a reality separate from the body, it cannot be multiplied or distinguished between individuals.

17- If the possible intellect were multiplied, the intelligible forms would also be multiplied and become individual, thus intelligible only in potential and not in act, which is inadmissible.

18- It is argued that what is common between the agent and the patient is essential. However, since the possible intellect shares nothing in common with phantoms (sensible images), it cannot be the same intellect that we possess internally and therefore would not multiply between persons.

19- Everything that exists as one does not depend on another to be so. Since the possible intellect does not depend on the body to exist, its unity also does not depend on the body and, consequently, cannot be multiplied with bodies.

20- Aristotle teaches that in pure forms, essence is the same as species. Since the possible intellect is a pure form, if the nature of the species is one, the intellect is also one in all intellectual animals.

21- The multiplication of souls according to bodies occurs only by virtue of their union with them. Since the possible intellect is understood as that which transcends union with the body, it does not multiply among humans.

22- If the intellect were dependent on the multiplication of the body, intelligible species would multiply, which contradicts their nature as intelligible forms in act. Therefore, neither the soul nor the possible intellect can multiply.

This series of arguments constitutes a defense of the idea that the possible intellect is unique for all humans, as its separate and universal nature prevents it from multiplying or distinguishing between individuals.

> Next, St. Thomas presents two arguments from authority asserting that the possible intellect is not the same for all humans

1. First argument against the uniqueness of the possible intellect. If the possible intellect were unique and common to all humans, then what one person understands or knows would also be understood or known by any other person. This is obviously false, because each person has different intellectual knowledge and experiences. Therefore, this argument suggests that there must be a distinction in the possible intellect for each individual, allowing each person to have their own knowledge.

2. Second argument based on the relationship between the intellectual soul and the body. The intellectual soul (or possible intellect) is related to the body in two ways: as form, which gives existence and organization to the body, and as motor, which guides and moves the body as its instrument. Following this logic, each form requires a specific matter, and each motor requires a specific instrument. Thus, it would be impossible for one intellectual soul to function in multiple bodies, since each body would need its own form or soul adapted to its individual characteristics.

These arguments support the idea that each human being has a distinct and separate possible intellect, as sharing a single intellect among all individuals would not be consistent with the individuality of intellectual experience and the specific relationship the soul has with the body.

> Next, St. Thomas provides his own response to the Question raised

The Angelic Doctor explores the question of whether the possible intellect, the faculty that enables the capacity to know in humans, is a unique and common entity for all or if it multiplies in each individual. The analysis focuses on ontological and epistemological arguments regarding the nature of this intellect. Here are the main points of his reasoning:

1- Dependence of the intellect on the body. St. Thomas asserts that if the possible intellect exists as a substance separate from the body, then it should be one for all, as separate substances do not multiply by the variety of bodies. However, this conclusion presents important problems, especially regarding how each individual can have different knowledge if they all shared the same intellect.

2- Special difficulty in the unity of the intellect. St. Thomas points out that it seems absurd for everyone to share the same intellect, as knowledge varies from person to person. This would be impossible if the intellect were one and the same in all, as a common perfection cannot be the basis for a diversity of knowledge in each individual.

3- Argument from phantasms *(phantasmata)*. Some philosophers try to solve this problem by claiming that "intelligible species" are in the particular phantasms of each person, not in the common intellect. Thus, although the intellect is one, knowledge is different due to the diversity of phantasms. St. Thomas rejects this idea because he considers that species are not intelligible in act until the intellect abstracts them from the phantasms, so there cannot be diversity in knowledge simply by having different phantasms.

4- Operation of the intellect in different individuals. St. Thomas argues that if the intellect were unique, then the operation of "understanding" would also be unique and, therefore, could not be attributed to particular individuals. Furthermore, it would be impossible for two people to simultaneously understand the same concept at the same moment. This shows a contradiction in the idea of a common intellect, as intelligence is a particular operation in each person.

5- Impossibility of a unique intellect. He concludes that the intellect cannot be one in all, as this would contradict the multiplicity of experiences and knowledge. St. Thomas asserts that the intellect must multiply with each human soul, and being part of human nature, it is necessarily tied to each individual in particular.

6- Nature of the soul and multiplicity. St. Thomas explains that the human intellect individualizes in each person as a property of the soul, which multiplies with each human body, similar to how certain physical qualities can be the same in essence but different in each individual.

In summary, St. Thomas argues that the possible intellect is not a unique entity shared by all humans but is individualized in each person. This individualization allows each human being to have their own knowledge and particular experiences, in line with the idea that each human soul is unique and has its own intellect.

> Next, St. Thomas responds to each of the twenty-two arguments initially presented, according to which there is unity between the possible intellect and the intellectual soul in all men

1- Response to the first argument. St. Thomas responds that truth is the adequacy of the intellect to the thing (reality). Thus, when different people know the same truth, it is because their conceptions align with the same reality.

2- Response to the second argument. It is clarified that St. Augustine would be deemed ridiculous not for claiming that many souls exist, but for saying that they are many both in number and in species, which would imply an unjustified duplicity.

3- Response to the third argument. St. Thomas explains that the possible intellect is not multiplied due to any difference in form, but by the multiplication of the very substance of the soul, of which it is a potency.

4- Response to the fourth argument. It is not necessary for the common intellect to be separated from what it knows; only the intellect in potency must be free from the nature of what it receives. Therefore, an intellect that is already in act (like the divine intellect) knows itself inherently, while the possible intellect knows itself through the intelligible species of other objects.

5- Response to the fifth argument. It is clarified that the possible intellect has nothing in common with sensible natures, from which it receives its intelligibles, although one possible intellect is specifically the same as another.

6- Response to the sixth argument. St. Thomas asserts that in beings separated from matter, distinction can only occur according to species, and different species are configured in degrees, just as numbers are diversified by addition or subtraction. However, multiplication in separated beings is not acceptable in the Christian faith.

7- Response to the seventh argument. Even though several individuals possess the same intelligible species in their respective intellects, what is understood through these species is one and the same, since the object of universal knowledge is identical in all cases. This unity is due to the immateriality of the intelligible species.

8- Response to the eighth argument. Platonists argue that the fact that something is one in many comes from the thing itself. Thus, they argue for the necessity of ideas as participation in natural things and in universal intelligences. For Aristotle, however, the understanding of one in many comes from the abstraction of the intellect, which abstracts from the principles of individuation.

9- Response to the ninth argument. It is explained that the intellect is a "place" of the species because it contains them, but this does not imply that the possible intellect is one for all men, rather it is common to all species.

10- Response to the tenth argument. Unlike the intellect, the sense cannot be considered a place of the species because it requires an organ to receive them.

11- Response to the eleventh argument. St. Thomas clarifies that the possible intellect "operates everywhere" not because its operation is everywhere, but because it relates to things that are everywhere.

12- Response to the twelfth argument. Although the possible intellect does not have a determined matter, the substance of the soul, of which it is a potency, does have it, not in the sense of being from it, but in the sense of being in it.

13- Response to the thirteenth argument. The principles of individuation do not belong to the essence of forms, but this only applies in the case of substances composed of matter and form.

14- Response to the fourteenth argument. A distinction is made between the first mover of the heavens, absolutely separate from matter, and the human soul, which is not similar in its relationship to matter.

15- Response to the fifteenth argument. Separated souls do not differ in species but in number, because they can unite with specific bodies.

16- Response to the sixteenth argument. Although the possible intellect is separated from the body in terms of its operation, it is a potency of the soul, which is the act of the body.

17- Response to the seventeenth argument. Something is understood in potency not by being individual, but by being material. Thus, intelligible species, although individualized, are understood in act by the intellect.

18- Response to the eighteenth argument. The phantasm moves the intellect by becoming intelligible in act, through the action of the agent intellect, to which the possible intellect is related as a potency with respect to an agent.

19- Response to the nineteenth argument. Although the being of the intellectual soul does not depend on the body, it has a natural inclination toward it for the perfection of its species.

20- Response to the twentieth argument. While the human soul does not include matter as part of itself, it is the form of the body, and its essence includes the relationship with the body.

21- Response to the twenty-first argument. Although the possible intellect is elevated above the body, it is not elevated above the entire substance of the soul, which is multiplied in relation to different bodies.

22- Response to the twenty-second argument. St. Thomas seeks to clarify that the relationship of the soul with the body does not mean that everything in the essence of the soul is subject to materiality. **St. Thomas asserts that, although the soul is united to the body as its form, not all its operations depend on matter**. In this sense, it is reaffirmed that there are aspects of the soul's nature (particularly, the intellection of essences) that transcend the material, given that the act of intellection does not operate in an organic way nor depend on a physical organ.

4. QUESTION 4: Whether it is necessary to admit the existence of an agent intellect

> St. Thomas presents nine arguments in which various authors assert that it does not seem necessary to affirm the existence of an agent intellect

1- What can be accomplished by a single means in nature should not be done by several. The human being can understand adequately through a single intellect, which is the possible intellect. Therefore, it is not necessary to postulate an agent intellect.

If the possible intellect is sufficient for human understanding, then there is no need to assume the existence of an agent intellect. This implies that each power of the soul can operate autonomously without requiring an external agent to complete its function.

2- The sense of touch and sight are different powers but can influence each other. For example, a blind person can imagine something they have not seen based on their sense of touch. This shows that both powers are connected to the same essence of the soul. Therefore, if the possible intellect is a power of the soul, the imagination can also influence the intellect. Thus, there is no need for an agent intellect.

This argument highlights the interconnectedness of the powers of the soul, suggesting that the possible intellect can receive influences from the imagination without requiring an agent intellect to mediate.

3- The agent intellect is postulated to convert intelligibles in potency into intelligibles in act. However, the possible intellect can receive ideas without the need for an agent intellect, since it has the capacity to receive according to its immaterial nature. Therefore, there is no need for an agent intellect.

The possible intellect, being immaterial, can receive knowledge without requiring the intervention of an agent intellect, reinforcing the idea that it is unnecessary to postulate one.

4- Aristotle compares the agent intellect to light. Light is not essential for seeing unless it makes the medium (like air) visible. In the same way, the agent intellect is not necessary for the possible intellect to be prepared to receive knowledge because the latter already has the capacity to do so.

Here, it is argued that the possible intellect innately possesses the ability to understand, similar to how colors are visible without the need for an agent to illuminate them.

5- Just as the intellect relates to intelligible things, the senses relate to sensible things. Sensible objects can move the senses without the need for an external agent. Therefore, intelligibles do not require the intervention of an agent intellect.

A parallel is drawn between sensory powers and intellectual powers, indicating that both operate independently and do not require an external agent to function.

6- For something in potency to become actual, it is enough for there to be something actual of the same nature. For the intellect in potency to become actual, only an intellect in act is needed, which can be the same one that understands.

This argument emphasizes that the process of acquiring knowledge can be achieved through direct experience or teaching, eliminating the need for an agent intellect to facilitate it.

7- The agent intellect is proposed to illuminate our mental images, just as sunlight illuminates colors. However, divine light is sufficient to illuminate our understanding, so there is no need to postulate an agent intellect.

Here, it is emphasized that knowledge and understanding can come from a higher source (divine light) and do not require an agent intellect to be illuminated.

8- If there are two types of intellect, the agent and the possible, this would imply that one person would have two ways of understanding, which seems inconvenient.

Having two intellects functioning separately within the same person would complicate understanding and be impractical.

9- The intelligible species is considered a perfection of the intellect. If there were both an agent intellect and a possible intellect, there would be a duplicity in the perfection of understanding, which would be unnecessary.

It is suggested that the existence of two intellects would result in an excess of perfection, which would not be necessary or useful for human understanding.

Through these arguments, it is defended that the possible intellect is sufficient to carry out the process of comprehension or understanding, without the need for an agent intellect. It is affirmed that the powers of the soul can interact effectively without the intervention of an external agent.

Next, St. Thomas presents an argument from authority asserting the necessity of an agent intellect

An objection (or "sed contra") is raised to the position that it is not necessary to postulate the existence of an agent intellect. The reference to Aristotle in his work *De Anima* is used to support this objection.

The explanation of the text can be broken down as follows:

1- Principle of action in nature. Aristotle argues that in all of nature there are two fundamental aspects: what acts (the agent) and what can be acted upon (the potential). That is, in any process of change or in any being, there is always something that causes the change and something that receives that change.

2- Application to the nature of the soul. When applying this principle to the discussion about the soul, it is proposed that just as in nature in general there are these two aspects, in the human soul there must also be. This implies that there must be a distinction between the intellect that acts *(intellectus agens)* and the intellect that receives *(intellectus possibilis)*.

3- Function of the agent intellect. The agent intellect is what allows the abstraction of intelligible species from sensible things. The possible intellect, in turn, is what receives these species and formulates the universal concept. Without the existence of both, it would be difficult to explain how the process of understanding and knowledge occurs in humans.

In summary, this argument holds that, following Aristotelian logic, it is essential to recognize the existence of an agent intellect in the soul, as it is necessary for the process of comprehension and knowledge, in which there is an agent that acts and a potential that receives that action.

> Next, St. Thomas offers his own response to the raised Question

Indeed, the Angelic Doctor defends the necessity of postulating the existence of an agent intellect *(intellectus agens)* to explain how the process of knowledge functions. The main ideas he develops are explained below:

1- Necessity of the agent intellect. Thomas argues that it is necessary to postulate an agent intellect because the possible intellect *(intellectus possibilis)* is in potency with respect to the ideas or concepts it must understand (the "intelligibles"). This means that the possible intellect can receive knowledge, but it needs to be moved by something that is already intelligible.

2- Movement of the possible intellect. For the possible intellect to move (i.e., to understand something), there must be an object that moves it. However, the objects understood by the possible intellect do not exist in nature as independent entities, because the intellect does not understand things in their individuality, but understands them as universals (as a common idea that can apply to several individuals).

3- Abstraction of matter. Thomas argues that the agent intellect performs the task of abstracting the ideas from the material conditions that individualize them. For example, the nature of a species has no intrinsic reason to multiply into different individuals, and the principles that individualize it are external to its own reason. Thus, the intellect can grasp the essence of things (i.e., what makes them what they are) without being limited by individual particularities.

4- Contrast with the Platonists. Thomas mentions that if universals existed in reality by themselves, as the Platonists claimed (through Ideas or Forms), there would be no need for an agent intellect. In that case, material objects would move the possible intellect directly, without the mediation of an intellect that abstracts. However, since St. Thomas disagrees with the Platonic theory of Ideas, he considers it necessary to postulate the existence of the agent intellect.

5- Knowledge of immaterial substances. Although there are some beings that are intelligible in themselves (such as immaterial substances), the possible intellect cannot directly access them. Instead, it comes to know these realities through abstraction from the material and sensible objects it encounters.

> Next, St. Thomas responds to each of the nine arguments initially presented, according to which it is unnecessary to affirm the need for an agent intellect

1- Response to the First Argument. St. Thomas points out that the human intellect cannot function solely with the possible intellect. The latter needs to be activated by something that is already intelligible, because ideas themselves do not exist in nature. Therefore, there must be an agent intellect that produces those intelligibles. Although there are different powers in the soul, their interaction is not sufficient for understanding without the intervention of the agent intellect. This intellect acts by moving the possible intellect in such a way that universal ideas can be formed from individual experiences.

2- Response to the Second Argument. At this point, it is argued that the imagination must first have forms in memory in order to form concepts related to vision. For example, a person born blind cannot imagine colors, since they lack prior sensory experiences that would allow them to form the idea of color. This indicates that the agent intellect is necessary to process and relate these ideas, as the possible intellect cannot act without this prior knowledge.

3- Response to the Third Argument. Thomas clarifies that the condition of the receiver cannot transfer a form from one type to another; however, it can vary within the same type according to how it manifests. This means that since universal and particular species are different, the possible intellect alone is not enough to transform particular forms of the imagination into universals. Therefore, the action of the agent intellect is required to achieve this universalization.

4- Response to the Fourth Argument. Here, the relationship between light and vision is discussed. Some philosophers held that light is necessary to see because it allows colors to become visible, while Aristotle argued that colors are visible by themselves. However, it is concluded that light is essential to make a medium transparent and allow the visibility of colors. Similarly, the agent intellect is necessary for concepts that are in potency to become actual concepts, which means that it is fundamental for the understanding of ideas.

5- Response to the Fifth Argument. It is maintained that a sensible object, being particular, cannot influence the perception of another type of form; that is, the possible intellect can receive universal ideas that are not contained solely in the particular forms of the imagination. This distinction highlights that the agent intellect is essential for the understanding of intelligibles, as opposed to the sensory capacities that only handle the particular.

6- Response to the Sixth Argument. It is argued that the possible intellect, even though it is in act, cannot produce knowledge without the agent intellect. In the process of learning, the possible intellect may be partially in act and in potency. However, in order for knowledge of the principles to be acquired, the intervention of the agent intellect is needed, which acts as a mediator in the learning process. This intellect allows knowledge to develop from sensory experiences.

7- Response to the Seventh Argument. Just as in the natural realm there are active principles specific to each genus, an intellectual "light" specific to human beings is also required, in addition to the divine influence that acts as the general cause of the illumination of understanding. This reinforces the idea that, although God is the source of all light and knowledge, there is also a need for an agent intellect that operates within each individual.

8- Response to the Eighth Argument. It is clarified that although there are two types of intellect (the possible and the agent), this does not imply that there are two forms of understanding in humans. Both actions—receiving intelligibles and abstracting intelligibles—must work together for understanding to occur. In other words, knowledge occurs when both intellects collaborate.

9- Response to the Ninth Argument. It is explained that the intelligible species relates to both the agent and the possible intellect, but in different ways. While the possible intellect passively receives the forms, the agent intellect actively creates these forms through the process of abstraction. This emphasizes the active role of the agent intellect in the acquisition of knowledge.

Through these arguments, Saint Thomas Aquinas establishes the necessity of the agent intellect for the process of knowledge, highlighting the interaction between the powers of the soul and how active action is required to transform knowledge into actuality.

5. QUESTION 5: Whether there is one separately existing agent intellect for all men

St. Thomas presents ten arguments suggesting that the agent intellect is unique and separate

1- *The Philosopher* notes that the agent intellect is always active and never experiences moments of inactivity. In human experience, however, everything has periods of activity and inactivity. This suggests that the agent intellect must be a separate entity, not bound by human experience, implying that it is unique and universal.

2- It is stated that something cannot simultaneously be in potentiality and actuality in relation to the same thing. Since the possible intellect is in potentiality with respect to all intelligibles and the agent intellect is in actuality with respect to them, it seems incompatible for both to reside in the same substance of the soul. Therefore, the agent intellect must be separate from the possible intellect.

3- The possible intellect is in potentiality with respect to intelligibles, meaning it does not yet possess them actively. In contrast, the agent intellect acts on these intelligibles, making them accessible and understandable. It is argued that the possible intellect cannot be in actuality with respect to the intelligibles it already possesses, as if it were, it could not be considered potential but would instead represent knowledge already acquired.

The text relates to the idea that the agent intellect is one and separate, emphasizing its unique, active role in the process of knowledge. This uniqueness is essential for abstraction, allowing human understanding to access universal ideas independently of individual particularities.

4- *The Philosopher* attributes characteristics to the agent intellect that seem specific to separate substances, such as perpetuity and incorruptibility. This suggests that the agent intellect is indeed a separate substance.

5- It is argued that the intellect does not depend on bodily constitution, and although understanding capacity varies among people due to bodily differences, this does not imply that the agent intellect is part of our constitution. Therefore, the agent intellect appears to be something separate from our nature.

6- It is stated that only one agent and one patient are needed for any action. If the possible intellect is part of our substance and the agent intellect as well, it would seem that we have everything necessary to understand. However, it is established that we actually need the senses, teaching, and divine illumination, indicating that the agent intellect cannot merely be something we possess.

7- The agent intellect is compared to light, suggesting that just as sunlight can make all that is visible visible, a single, separate light could suffice to make all intelligible things intelligible. This would imply that an agent intellect within us is unnecessary.

8- It is argued that the agent intellect resembles art, and art is a principle separate from its object. Therefore, it is concluded that the agent intellect must also be a separate principle.

9- It is asserted that the perfection of any nature implies that it resembles its agent. If the agent intellect were part of our soul, the perfection of the soul would depend on something within it, which would be absurd, as it would imply that the soul could find fulfillment within itself. Therefore, the agent intellect cannot be something that belongs to us.

10- It is stated that the agent is more noble than the patient, as mentioned in book III of *De Anima*. If it is granted that the possible intellect is in some way separate, then the agent intellect, which acts upon the possible, should be even more separate. This implies that the agent intellect cannot reside within the substance of the soul but must be entirely outside of it. This reinforces the idea that the agent intellect is a separate entity transcending the existence of the human soul.

> Following this, St. Thomas presents two arguments from authority, suggesting that the agent intellect is not a separate entity from the soul

1- As stated in Book V of *De Anima*, in every nature there is a fundamental distinction between two aspects: **the passive** or potential and **the active**, which configures and actualizes that potential. In the case of the soul, it is necessary to recognize these differences, one of which refers to the possible intellect (passive aspect) and the other to the agent intellect (active aspect). Therefore, both the possible intellect and the agent intellect are components that belong to the essence of the soul and cannot be considered separate entities. This distinction helps to understand how the human intellect transitions from the potential to know (possible intellect) to the active state of comprehending or effectively understanding (agent intellect).

2- Additionally, it is argued that the operation of the agent intellect consists in abstracting intelligible species from the images (or phantasms) we have in our mind. This abstraction always occurs within us, and if the agent intellect were a separate substance, there would be no reason for this abstraction to happen at certain times and not at others. The regularity of this function suggests that the agent intellect must be closely related to the soul and cannot be considered a separate entity.

> Following this, St. Thomas offers his own response to the Question posed

The text presents St. Thomas's argument on the nature of the agent intellect and its relationship to the possible intellect, as well as his position on separate entities and God. We break it down into nine points for clarity, as follows:

1- Nature of the agent and possible intellects. St. Thomas holds that the agent intellect is **more appropriate** to be considered a separate entity than the possible intellect. The possible intellect manifests in two states: sometimes in potentiality and other times in actuality. In contrast, the agent intellect enables true understanding, that is, it performs the act of understanding.

St. Thomas considers the agent intellect **more appropriate** as a separate entity due to its active and universal nature, which allows it to operate independently of the material and particular limitations of the possible intellect, which is closely tied to the human essence and sensory experiences. This distinction highlights the complexity of human understanding and its relationship to higher or universal principles of knowledge.

2- **Differentiation between acts and potencies**. The agent intellect, as an active principle, can be separate from what it brings into action, whereas the possible intellect, which is an internal capacity within humans to understand, must be intrinsic to the essence of the being.

Acts: This term refers to the actual execution of an action or the state in which an entity operates actively. In this context, the agent intellect is an active principle that performs the action of abstracting.

Potencies: In contrast, potency refers to the capability or possibility of performing an action. In the case of the possible intellect, it is defined as the internal capacity of humans to understand, meaning their potential to receive and process information.

The agent intellect is considered an active principle because it is responsible for carrying out intellectual acts. It can be "separate from what brings it into action," implying that the agent intellect can function independently or abstractly from sensory images and experiences. This means it can abstract ideas and concepts without needing to be immediately connected to specific sensory data. This capacity to abstract suggests a form of independence and a higher level of intellectual activity.[2]

On the other hand, the possible intellect is defined as an internal capacity that is intrinsically linked to the essence of the human being. This implies that the possible intellect is closely tied to human nature itself; it cannot be separated from it without losing its essence. It is the aspect of the mind that receives and comprehends ideas presented through sensory experiences and mental images (phantasms).

The differentiation between these two types of intellect highlights the complexity of the human cognitive process. While the agent intellect can operate more abstractly and actively, the possible intellect is limited to being a passive capacity that requires sensory information to function. Both are essential to understanding, but they play different roles in the dynamics of knowledge.

The agent intellect can act independently and has an active role in generating knowledge, whereas the possible intellect is an internal and essential capacity that must be present in humans to enable comprehension. This distinction is fundamental for understanding how the human mind functions according to the philosophy of St. Thomas Aquinas.

3- Concept of separate intelligences and entities. Some philosophers argue that the agent intellect is a separate substance, which they refer to as an "intelligence." This intelligence relates to human souls similarly to how higher substances relate to the souls of celestial bodies.

4- The relationship between god and the agent intellect. Catholic faith teaches that God alone acts in our souls, not some other separate substance. Some have argued that the agent intellect might be God, the "true light" that enlightens every person. However, St. Thomas considers this position inadequate.

5- Universal and particular active principles. Just as celestial bodies serve as universal active principles for lower bodies, a particular active principle is required for the operations of living beings, which, in the case of humans, is the agent intellect.

This text addresses the relationship between active principles in Thomistic philosophy, especially concerning celestial bodies, lower bodies, and human intellectual operation.

St. Thomas posits that celestial bodies (such as stars and planets) act as universal active principles. This means they have a general impact on all lower bodies, like those on Earth. Their influence is observable in natural phenomena such as climate, tides, and other aspects of the physical world. These principles are "universal" because they affect a wide range of entities or phenomena indiscriminately.

On the other hand, among living beings, a particular active principle is required to influence their specific operations. Unlike universal principles that have a general effect, particular active principles apply to concrete situations or entities. For humans, this particular active principle is the "agent intellect."

The agent intellect is a human capacity that enables abstraction and understanding of ideas. It is the aspect of our mind that transforms sensory images and experiences into knowledge. Unlike universal active principles that affect all bodies equally, the agent intellect operates on an individual level, allowing each human being to perform specific acts of knowledge and understanding.

The distinction between universal and particular active principles highlights the complexity of action and knowledge. While celestial bodies influence the world generally, the agent intellect is essential for intellectual activities and human knowledge development. This particular principle is fundamental to human nature, enabling us to not only receive information from the outside world but also actively process and understand it.

6- Implications of a separate agent intellect. If it is claimed that the agent intellect is an entity separate from God, this would imply that the ultimate perfection and happiness of humans would depend on union with something other than God, which contradicts the Gospel's teaching on eternal life as the knowledge of God.

7. The difficulty of a separate agent intellect. St. Thomas also argues that it is impossible for the agent intellect to be a separate substance for the same reasons that apply to the possible intellect. The operations of the agent intellect (abstraction) and the possible intellect (reception of intelligible forms) are experienced within us. Each operation requires an intrinsic formal principle that cannot be merely external.

8- Interaction of potency and act. In human nature, mental images (phantasms) can be seen as being in potency concerning the entities they represent, while they are in act as likenesses of specific things. The possible intellect is in potency to all intelligibles, but it becomes determined to understand through abstract species.

9- Activity of the agent intellect. The agent intellect is presented as an active power that abstracts images from their material conditions, similar to light making colors visible. The notion that the agent intellect is merely a habit of indemonstrable principles is refuted, as we also abstract these truths from the singular, indicating that the agent intellect must exist as the cause of principles.

Thomas Aquinas argues that both the possible intellect and the agent intellect are essential to the nature of human understanding, and that both reside within the soul rather than existing as separate entities, as this would lead to theological and philosophical confusions that contradict the Catholic faith.

> St. Thomas then responds to each of the ten arguments initially posited that viewed the agent intellect as a separate entity

1- Thomas clarifies that the statement by *The Philosopher* about the agent intellect and the intellect in act does not apply to the agent intellect but rather to the intellect in act. He explains that, according to Aristotle, it is necessary to distinguish between the possible intellect and the intellect in act. The possible intellect and the object understood are not the same, while the intellect in act is identical to the object understood in act. He also notes that, although the possible intellect may sometimes understand and sometimes not, the intellect in act is always in a state of understanding.

2- It is noted that the substance of the soul is in potency and act with respect to the same phantasms (mental images), but not in the same way. This suggests that the agent intellect acts differently in relation to images than the possible intellect does.

3- The possible intellect is in potency regarding intelligibles, according to their existence in phantasms. In contrast, the agent intellect acts on these intelligibles but in a different manner, as shown previously.

4- It is clarified that *The Philosopher's* statements about what is separate and immortal do not refer to the agent intellect, as it was previously mentioned that the possible intellect is also separate. These statements should be understood in the context of the intellect in act, which encompasses both the agent and possible intellects. In this sense, only the intellect that encompasses both aspects is separate, immortal, and perpetual, since other parts of the soul do not exist without the body.

5- Differences in bodily constitution cause variations in the capacity to understand, which depends on the powers the intellect uses, such as imagination and memory, which require bodily organs.

6- Although the human soul contains both an agent and a possible intellect, something external is required to enable understanding. Phantasms derived from sensory experiences, which represent the likenesses of things to the intellect, must exist. The agent intellect alone is insufficient to grasp specific forms without the assistance of external elements guiding it. Moreover, if we consider the agent intellect as a shared virtue within our souls, an external cause from which that light derives is necessary, and this cause is identified as God, who provides an additional understanding that transcends natural reason.

7- Colors that affect sight are external, while phantasms that activate the possible intellect are intrinsic to the human being. Thus, although sunlight is sufficient to make colors visible, for phantasms to be intelligible in act, the inner light of the agent intellect is needed. Furthermore, it is argued that the intellectual part of the soul is more perfect than the sensitive part, which justifies the need for more complete principles to enable its operation.

8- Thomas acknowledges a certain similarity between the agent intellect and art but clarifies that this comparison should not be extended in all respects.

9- The agent intellect alone cannot bring the possible intellect to a state of complete perfection, as it does not contain the specific forms of all things. Therefore, the possible intellect must be joined to something that holds the forms of all things, and this ultimate source is God.

10- Finally, it is established that the agent intellect is nobler than the possible intellect, just as active power is nobler than passive power. This distinction is justified by the greater separation of the agent intellect from matter; however, this does not imply that it is a completely separate substance.

6. QUESTION 6: Whether the soul is composed of matter and form

> St. Thomas presents seventeen arguments by which it seems that the soul is composed of matter and form

1- Boethius' argument on simplicity and being a subject: According to Boethius in the *Book on the Trinity*, a simple form cannot be a subject. Since the soul is the subject of sciences and virtues, it cannot be a simple form; therefore, it must be composed of matter and form.

2- Participation and non-participation in being: Boethius in the *Book of the Hebdomads* establishes that something that "is" can participate in something, but being itself does not participate. Since the soul participates in qualities that inform it, it cannot be only a form, but should have matter.

3- Act and potency in the being of the soul: The soul is considered in philosophy as "form," meaning it is the principle that gives life and activity to a being. However, if the soul were only "form," it could not exist by itself, as it would be in a state of potentiality, meaning it would lack something to fully realize its existence. In this context, it is said that each type of potentiality corresponds to a unique act; therefore, if the soul were only form, it could not be the subject of another action or substance.

Despite this, it is evident that the soul acts as a subject, meaning it can perform actions and be influenced by its environment. This suggests that the soul is not a simple substance; rather, it is a combination of matter and form. This combination allows the soul to interact with the world and receive external influences. In summary, the soul is a composite entity that can act and also be affected by what surrounds it.

4- Specific and individual accidents: Material accidents correspond to individuals, while formal accidents correspond to complete species. Since the soul has individual accidents (such as musical ability), it cannot be only form and should be composed of matter and form.

5- Principle of action and passion: Form is the principle of action, and matter of passivity. Since there is both action and passion in the soul (e.g., in both passive and active understanding), it must be composed of matter and form.

6- Properties of matter in the soul: Characteristics such as being in potency, receiving, or underlying are material properties, which are also observed in the soul. This suggests that the soul has matter.

7- Common agents and patients: According to ancient philosophy, agents and patients must share a common matter. Since the soul can suffer from material things (such as the fire of hell according to St. Augustine), it must have some matter.

8- Divine action on the soul: Every action ends in a composite of matter and form, as Aristotle states in the *Metaphysics*. If God acts on the soul, then the soul must be a composite of matter and form.

9. Dependence of the soul on God for being and unity: Something that is only form would automatically be a being and a unity in itself. However, the soul requires God for both its being and its unity, which would indicate a composition of matter and form.

10- Reduction of potency to act: Everything that passes from potency to act must be composed of matter and form. Since the soul needs an efficient cause to move it from potency to act, it must have matter.

11- Reference to Alexander of Aphrodisias on intellect: This text refers to the interpretation of Alexander of Aphrodisias on the nature of intellect in the soul. Alexander, a Peripatetic philosopher, following Aristotle's teachings, introduces the idea of "hylomorphic intellect" to explain how the human intellect works. The term " hylomorphic " combines two Greek terms: *hylé* (matter) and *morphé* (form). Thus, the concept of "hylemorphic intellect" implies that the intellect has an aspect of "prime matter."

In Aristotelian philosophy, prime matter is the potential substrate of all forms and lacks intrinsic characteristics until a form actualizes it. Applying this to the intellect, Alexander suggests that the intellect has a receptive capacity, as if it were "matter" in the sense of being passive and able to receive forms of knowledge.

If the intellect is hylomorphic, then, according to Alexander, there is a kind of "matter" in the soul in a philosophical sense. This does not mean that the soul is made of physical matter, but rather that it has a disposition to receive knowledge and concepts, like a "potency" that is actualized when it acquires intellectual forms (ideas).

The reference to Alexander suggests that the soul is not a pure, unchanging form but has a potentiality similar to matter. This would imply a certain "composition" in the soul, as it is considered partially passive and potential, a characteristic typically associated with matter in Aristotelian philosophy.

12- Composition of potency and act in the soul: Every being is either pure act, pure potency, or a composite of both. Since the soul is neither pure act (exclusive to God) nor pure potency (proper to matter), it must be a composite of act and potency.

13- Individuation and matter: Individuation depends on matter. Since the soul is individuated, it must possess some kind of matter.

14- Suffering of the soul due to sensitivities: Since the soul experiences passions from sensible and material things, it seems to have something in common with the material, which indicates a material composition.

15- Classification of the soul as a species: Like angels, the soul is considered a species within a genus. This implies a composition of matter and form, as the genus acts as matter and the specific difference as form.

16- Diversification of common forms through matter: Intellectuality is a common form in souls and angels. For this common form to be distributed across many individuals, there must be a divisive matter in each case.

17- Movement and matter in the soul: Everything capable of movement has matter. St. Augustine asserts that the soul is subject to change, which means it cannot be of divine nature and must be composed of matter and form.

Each of these arguments seeks to support the idea that the human soul has a structure composed of matter and form, and questions the notion that it is a simple and separate form.

> Next, St. Thomas presents an argument from authority that objects to the idea that the soul is composed of matter and form

The argument is developed as follows:

1- Initial premise: Everything composed of matter and form has a form. This means that if something is composed of matter and form, that combination requires a specific form that gives it its identity.

2- Hypothesis: If the soul is composed of matter and form, then it must have an additional form that gives it its particular configuration.

3- The problem raised: However, the soul itself is already considered a form, not something that requires an additional form. In Scholastic philosophy, the form is the principle that gives life and configures beings. If the soul (which is a form) needed another form, this would lead to a paradox, because then each form would need another form to configure it, and so on.

4-. Conclusion: This reasoning would lead to an infinite regression *(ire in infinitum)*, something considered problematic and impossible in

Scholastic philosophy. Therefore, if we accept that the soul is a form, we cannot maintain that it is a composition of matter and form without falling into logical contradictions.

The objection concludes that the soul cannot be composed of matter and form, as it would be incoherent to give an additional form to what is already a form in itself.

> St. Thomas then provides his own response to the Question posed

The Angelic Doctor rejects the idea that the soul is composed of matter and form, a view defended by philosophers such as Avicebron.

He begins by mentioning that there are different opinions on this matter. Some think that all substances, except God, are composed of matter and form, following the thought of Avicebron, author of *Fons Vitae*.

Avicebron argues that any entity that possesses the properties of matter (such as the ability to receive and be the subject of something) must contain matter. Since the soul has properties similar to those of matter—being receptive and potential—Avicebron deduces that it must also be composed of matter. However, St. Thomas considers this idea frivolous and impossible.

St. Thomas' critique:

1- Differences in the way of receiving: St. Thomas explains that the act of "receiving" or "suffering" is different in the soul and in matter. Prime matter (matter without form) receives with a change or movement; however, the soul receives knowledge without undergoing physical change, that is, without movement.

2- The immaterial nature of the soul: Matter is present only in corporeal beings, as they have a physical location. In contrast, the soul receives without physical transformation, as Aristotle mentions in *De*

Anima, where the reception of knowledge occurs without suffering, as in physical bodies.

3- Inconsistency in the composition of the soul: For St. Thomas, if the soul were composed of matter and form, it would create a separate species in nature, independent of the body. But this contradicts the Aristotelian doctrine that the body and soul together form the human species, as the body is an essential part of this species.

4- Incompatibility of union with the body: If the soul were a composition of matter and form, it could not be the formal principle that gives existence to the body; only a part of the soul would be. Thus, it would not be the complete form of the body, which is contradictory, since the soul is what gives life and form to the body.

St. Thomas also rejects some theories that proposed complex or mystical ideas to explain how the soul is united to the body. In his time, there were philosophers and thinkers who considered that the connection between the soul and the body occurred through a kind of "light" or intermediary force, sometimes called "cosmic light." According to this theory, different classes of souls (vegetative, sensitive, and rational) would unite to the body through different forms of this "light" or energy.

St. Thomas considers these ideas "fantastic," because, in his view, they unnecessarily complicate the process of union between the soul and the body. According to his thinking, it is not necessary to imagine an "external light" or cosmic intermediary to explain this relationship.

Instead of resorting to the idea of an external light, St. Thomas argues that the soul is united to the body directly and naturally, like act to potency.

St. Thomas addresses the composition of the soul in terms of act and potency. Thus, he explains how the soul can be an independent and active entity without needing material composition.

Unlike material beings that are composed of matter and form, the soul is a "subsistent form." This means that, although the soul does not have matter, it exists independently and can subsist without the body.

In the human soul, St. Thomas finds another type of composition: that of essence *(essentia)* and act of being or existing *(esse* or *actus essendi)*.

The essence of the soul is "what the soul is," its nature or "what." *Esse* is the act of existing, the "being" in itself that makes the soul actually in act.

St. Thomas says that in the human soul, essence acts as potency in relation to esse, which acts as act. This means that the essence of the soul, by itself, has the capacity to exist, but it only becomes a real existing being, and thus complete, when esse gives it act, that is, when the essence receives the act of being or existing.

For St. Thomas, this structure of act and potency in the soul allows him to explain how the human soul can exist without depending on a body. As a subsistent form, the soul has a "potency" or capacity (its essence) that, when united with *esse*, completes it, giving it existence and reality.

The soul is a subsistent form, which can have a composition of act and potency (essence and act of being or existing), but not of matter and form. The composition of act and potency is present in all created things, where the essence (potency) receives the act of being. However, the composition of matter and form is restricted to material beings.

> Next, Thomas Aquinas responds to each of the seventeen arguments initially presented, which considered the soul to be composed of matter and form

1- Rejection of Boethius' idea. Thomas Aquinas argues that Boethius speaks of a form that is completely simple, referring to the divine essence, which is a pure act and cannot be a subject (or receive) because it has no potentiality.

2- Subsistent forms. Unlike the divine essence, other simple forms like angels and the soul are subsistent and can be subjects, as they have some potency. This means they can receive and act according to their potentiality.

3- Comparison of essence and forms. Here, Thomas Aquinas points out that a form is not only compared to the act of being (esse) as potency to act, but can also be compared with another form as potency to act (for example, transparency with light).

If transparency existed as a separate form, it could receive not only the act of being, but also light. This also applies to subsistent forms like angels and souls, which are capable of receiving both the act of being and other perfections.

The more perfect these subsistent forms are, the less they need to participate in other forms to reach their perfection, as they possess more perfection in their own nature.

4- Individuality of the soul. Thomas Aquinas clarifies that human souls are individual forms within bodies. Therefore, they can have accidental properties according to their individuality, although these do not apply to the entire species.

5- Passion in the soul. The passion attributed to the possible intellect is not the same as the passion in matter. Thomas Aquinas distinguishes between reception in the intellect, which is immaterial, and natural action, which involves the imprinting of forms in matter.

This implies that action and passion in the soul do not lead to the conclusion that the soul is a composition of matter and form.

6- Receiving and underlying. Terms like "receiving" and "underlying" are applied to the soul differently than to prime matter. This suggests that it is incorrect to assume that the properties of matter also apply to the soul.

7- Suffering of the soul. Although hell (which is material) affects the soul, it does so in a non-material way. The suffering the soul experiences is spiritual and relates to divine justice.

8- Generative vs. creative action. The act of generating is limited to composites of matter and form, whereas creative action is not restricted by matter.

9- Causes of subsistent forms. Subsistent forms do not need a formal cause to be united in unity and existence, as they are forms in themselves. However, they do need an external cause to give them existence.

10- Agent in motion. An agent in motion converts something from potency to act. An agent without motion does not convert from potency to act, but gives existence to what, by nature, is in potency to be.

11- Hylomorphic intellect. Or possible intellect. Some call it material intellect. This type of intellect is not material, but it has similarities to matter, as it is in potency with respect to intelligible forms, just as matter is in potency with respect to sensible forms.

12- Composition of the soul. Although the soul is neither pure act nor pure potency, this does not imply that it is a combination of matter and form.

13- Individualization of the soul. The soul is not individualized by the matter from which it is made, but by its relation to the matter in which it is found.

14- Sense and connection. The sensitive soul does not suffer from the sensibles themselves, but from its connection to them. Feeling is a type of suffering that does not refer only to the soul, but also to the animated organ.

15- Category of the soul. The soul is not strictly classified within a genus as a species, but rather as a part of the human species, which implies that it is not a composition of matter and form.

16- Intelligibility and diversity. Intelligibility is not distributed as a form of species among many, as it is spiritual and immaterial. It diversifies according to the forms, either in different species or simply in different individuals.

17- Mutability of the soul and angels. Both the soul and angels are considered mutable spirits, as they can change according to their choice. However, this mutation refers to changes in their operations, not in their essence, which is immaterial and does not depend on matter to change.

7. QUESTION 7: Whether the angel and the soul are of different species

> St. Thomas presents nineteen arguments according to which it seems that the angel and the soul do not differ specifically

1- It is argued that entities that have the same natural operation are of the same species. Since both the soul and the angels perform the same operation, which is understanding, it is concluded that they are of the same species.

2- It is mentioned that the soul understands through discourse, while the angel understands without discourse, suggesting that it is not the same operation. However, it is answered that different types of operation do not necessarily imply different powers. Some things can be understood without discourse (like the First Principles) and others with discourse (like the conclusions). Therefore, they do not differ in species.

3- Understanding with and without discourse is compared to movement and rest, arguing that discourse is a type of movement of the understanding. However, being in motion and being at rest do not differ in species, so neither do the modes of understanding.

4- It is pointed out that angels understand things in the Word, in the same way that the souls of the blessed do, and that this knowledge is without discourse. Therefore, there is no difference between the soul and the angel in terms of how they understand.

5- It is suggested that not all angels are of the same species, even though they all understand without discourse. This implies that the mode of understanding (with or without discourse) does not cause a difference in species among intellectual entities.

6- The possibility is mentioned that some angels understand better than others. However, it is answered that understanding better or worse does not imply a difference in species, as it only reflects a degree of perfection in understanding.

7- It is observed that all human souls are of the same species, even though they do not all understand in the same way. Therefore, the capacity to understand with more or less perfection does not imply a difference in species among intellectual entities.

8- It is argued that the human soul understands through discourse, considering causes and effects, and that angels also do so. Therefore, there is no difference in understanding between them.

9- It is established that those who are perfected by the same perfections are of the same species. Since angels and souls are perfected by grace, glory, and charity, it is concluded that they are of the same species.

10- It is suggested that entities that share the same end are of the same species. Since angels and souls seek the same eternal beatitude, it is concluded that they are of the same species.

11- It is pointed out that if angels and souls were of different species, then the angel should be in a higher order than the soul. However, it is stated that there are no intermediaries between the human mind and God, so they cannot differ in species.

12- It is argued that the impression of the same image does not cause a difference in species. Since both the angel and the soul are images of God, they cannot differ in species.

13- It is said that if the angel and the soul have the same definition, then they are of the same species. Damascene is cited, who defines the angel as an incorporeal substance with characteristics that also apply to the human soul, so both are of the same species.

14- It is argued that those who coincide in the last difference are of the same species, as this difference is what constitutes the species. Since angels and souls share the nature of being intellectual entities, they do not differ in species.

15- It is mentioned that those who are not in a species cannot differ in species. Since the soul is not in a species but forms part of a species (united to the body, it forms the human species), it cannot differ from the angel.

16- This argument holds that definition is an essential attribute of species. According to Aristotelian philosophy, the definition of a species must include both the genus and the specific difference that characterizes the members of that species. However, both angels and souls are considered simple, that is, they are not composed of matter and form. This simplicity implies that they cannot be defined in terms of a composition that would allow identifying a genus and a specific difference.

Since they cannot be defined in this way, the argument concludes that they cannot differ in species. In other words, if there is no definitional basis that allows establishing clear differences between angels and souls, it must be concluded that they belong to the same species in a broader sense, as they share the same nature of being intellectual and simple entities. This reinforces the idea that, despite differences in their functions or properties, the underlying essence of both is common, which implies that they cannot be considered as different species.

17- This argument is based on the classification of species according to Aristotelian logic, which establishes that each species is constituted by a genus (the general category to which it belongs) and a specific difference (the characteristic that distinguishes the members of that species from those of other species within the same genus).

In the case of angels and souls, the argument holds that there is no distinct foundation on which these terms of genus and difference can be established. This means that, since both angels and souls are considered simple and immaterial, their essence cannot be analyzed or divided into genera and differences in the way that is done with other composed beings (such as material beings, composed of matter and form).

Since no genus and specific difference can be identified that differentiates angels from souls, it is concluded that they cannot be classified as different species. In other words, both share a common nature that prevents them from being considered as distinct species, as the lack of a basis on which to build that classification leads to the conclusion that they belong to the same species.

18- This argument focuses on the idea that, according to Aristotelian philosophy, entities that differ in species do so through contrary differences, meaning opposing characteristics that allow for a clear distinction between one species and another. For example, in the case of material beings, contrary differences can include qualities like "hot" and "cold" or "wet" and "dry."

In the context of immaterial entities, such as angels and souls, the argument maintains that there is no contrariety. This is because these entities do not have material properties that can be opposed to each other, as occurs in the physical world. Contrariety, being a fundamental principle in the classification of species in philosophy, implies that for two entities to be considered of different species, they must present characteristics that oppose each other.

Since angels and souls are considered simple and immaterial, they lack the necessary properties to establish contrary differences. Therefore, the argument concludes that, since there are no contrarities between them, angels and souls cannot differ in species. In summary, the absence of opposing differences in their nature means that they belong to the same category or species, as they cannot be classified according to the distinctions that apply to material entities.

19- It is suggested that angels and souls seem to differ primarily in that the angel does not unite with the body while the soul does. However, it is clarified that the body is considered matter for the soul, and matter does not define the species of the form. Therefore, angels and souls in no way differ in species.

> Next, Saint Thomas presents an argument from authority, according to which the angel and the soul differ specifically

The argument centers on the relationship between difference in species and number, particularly in the context of angels and souls.

The text begins by stating that things that do not differ in species, but only in number, cannot differ unless there is a distinction based on matter. This means that, if two beings are of the same species, the only way they can be considered distinct is through their matter, that is, through the physical or material characteristics that individualize them.

It is then argued that both angels and souls are immaterial and, therefore, do not have matter. This characteristic is crucial because it implies that matter cannot be used to differentiate between them. Matter is an essential element in establishing the difference between beings; if they lack matter, they cannot be distinguished in that way.

The argument continues by suggesting that if the angel and the soul do not differ in species, then, following the logic presented, they should also not differ in number. This is because numerical differentiation can only occur through matter, and since both lack it, the logical conclusion would be that they are one and the same being in numerical terms.

However, it is established that this conclusion is false. In reality, it is recognized that there are multiple angels and multiple souls, which shows that, even though they are immaterial, they are not identical in number. This reality contrasts with the assertion that they cannot differ in species.

Finally, the text concludes that, since the angel and the soul cannot be considered identical in number (as there are many of both), they must differ in species. This implies that, despite their immaterial nature, there are fundamental differences that distinguish them as different beings, each with its own essence.

> Next, Saint Thomas offers his own answer to the Question raised

The solution Saint Thomas offers to the question of whether the human soul and angels belong to the same species is developed as follows:

1- Reference to Origen. Saint Thomas begins by quoting Origen, who argued that all rational creatures were created equal by God, but their free will led some to draw closer to God and others to distance themselves, resulting in diversity in creation. Origen sought to avoid ancient heresies, proposing that the variability in creatures comes from their moral choices and relationship with God.

2- Critique of Origen's position. However, Saint Thomas critiques this view, noting that Origen focused too much on the individual good of each creature without considering the good of the whole. A good architect does not make all parts of a house equally valuable, but assigns different values to the parts according to their contribution to the whole. Similarly, God, as the architect of the universe, does not create everything equally, as that would lead to an imperfect universe.

3-. Differentiation of creatures. According to Saint Thomas, if the human soul and angels are of the same species, then the difference between them must be sought in form. Considering both are immaterial, the only difference they could have would be formal, which would indicate that they are not of the same species. He also points out that angels and souls cannot be said to be composed of matter and form, as this would imply that they have common matter, which is not plausible.

4- Nature of angels and souls. Saint Thomas rejects the idea that angels and souls are composed of matter and form, arguing that such an assertion would lead to confusion. Instead, he holds that the difference between angels and souls should be considered through their various perfections, that is, in how they relate to their principle of existence (God).

5- Difference in degrees of perfection. He mentions that in material substances, the diversity of species relates to the degrees of perfection in nature. As one moves from elements to animals, a progression in perfection is observed. However, in immaterial substances, the difference in species is measured in relation to their proximity to the first agent (God), with substances closer to God being more perfect.

6- The human soul as the lowest in the hierarchy. In this context, Saint Thomas asserts that the human soul occupies the lowest place in this hierarchy of perfection. Unlike angels, the human soul is potentially capable of understanding and acquiring knowledge through sensory experience, which implies that it needs a body to reach its fullness.

7- Conclusion about species. Finally, he concludes that, since angels and the human soul differ in degree of perfection and in relation to their principle, they cannot be of the same species. The implication is that each type of spiritual being (angels and souls) has its own essence that distinguishes them, thus reflecting the variety and order in divine creation.

> Next, St. Thomas responds to each of the nineteen arguments initially presented, which claimed that the angel and the soul differ in species

1- Thomas Aquinas establishes that the mode of understanding in both angels and souls is not of the same species. If the forms that are the principles of operation differ in species, their operations must also differ. For example, heating and cooling are different operations because they depend on different forms (heat and cold). The intelligible species that the souls use are abstracted from sensible images, while those of the angels are innate. This implies that human and angelic understanding are different in species. The difference in how they understand also results in angels being able to understand without discursive reasoning, while souls need a discursive process to reach the essence of things.

2- St. Thomas holds that intellectual souls understand through species abstracted from sensible images, and they understand both principles and conclusions. Therefore, concerning souls, it is the same knowledge specifically speaking. In contrast, angels understand without the process of abstraction from sensible images. Thus, the understanding of angels and souls does not belong to the same species.

3- In this response, it is explained that motion relates to the species of that to which it is directed. In the case of understanding, the angel understands without the need for a discursive process, while the soul does so through a discursive process. Therefore, the understanding of angels and souls does not belong to the same species.

4- St. Thomas maintains that the species of a thing is determined by its natural operation, not by those actions that arise from participation in a higher nature. He uses the example of iron and wood, which can burn when incandescent, to illustrate that, although both substances have a similar operation in that state, they belong to different species.

In relation to the vision in the Word, this operation occurs through divine light and transcends the natural capacities of both souls and angels. Therefore, it cannot be concluded that angels and souls are of the same species, as their operations and natures are distinct.

5- Here, it is argued that even among different angels, the intelligible species are not equivalent. As an intellectual substance ranks higher in the hierarchy and gets closer to God, its forms of knowledge become more elevated and powerful. Therefore, even though angels understand without the need for discursive reasoning, this does not mean they belong to the same species.

6- It is mentioned that "more" and "less" can be understood in two ways: one, in terms of matter that participates in the same form in different ways; and another, in relation to different degrees of perfection of forms. In the latter case, the diversity of degrees can differentiate species, just like colors vary in relation to light.

7- Although not all souls understand in the same way, all use species of the same nature, which come from sensible images. The inequality in understanding results from the diversity of sensory virtues and the disposition of bodies, which does not generate a difference in species.

8- St. Thomas distinguishes two modes of knowing something through another: a. Knowledge of a distinct knowledge: This involves reasoning from principles to conclusions, using a logical process. b. Knowing by the species itself: This refers to understanding directly the essence of the object, without the need for reasoning.

In the case of angels, they know causes and effects through their own essence, which is similar to that of their cause (God). This allows them to have intuitive and direct knowledge, without resorting to a discursive process. Thus, their mode of knowing is immediate and non-analytical, unlike humans.

9- The perfections granted to angels and souls come from participation in the divine nature. However, this coincidence in perfections does not imply that they are of the same species.

10- Here, it is argued that what has a single natural and proximate end is one in species, but eternal beatitude is a final and supernatural end, so the conclusion does not stand.

11- St. Thomas clarifies that Augustine does not maintain that there is nothing between our mind and God in terms of dignity and nature, but that our mind is justified and beatified directly by God. He uses the analogy of a soldier under the king to explain this relationship. *As if saying that a simple soldier is immediately under the king, not because others of higher rank are not under the king, but because none has authority over him except the king.*

12- St. Thomas points out that neither the soul nor the angel is the perfect image of God, only the Son is. Therefore, it is not required that they be of the same species.

13- In this response, it is noted that the definition that applies to the angel does not apply in the same way to the soul. The angel is an incorporeal substance, while the soul cannot be described in this way.

14- It is argued that those who believe that the soul and the angel are of the same species rely on a strong argument, but it is not conclusive. The reason is that the difference that defines their species must be higher both in the quality of their nature and in their definition.

That is, simply stating that both the soul and the angel are "intellectual" is not enough to classify them together, as they must have a more significant fundamental difference. The comparison with sensible beings illustrates this: if all brute animals were considered the same species just because they are sensitive, the deeper variations in their natures would be ignored.

Although the soul and the angel share intellectual capacity, there are more relevant differences that prevent them from belonging to the same species.

15- St. Thomas mentions that the soul is part of the species and, at the same time, a principle that gives the species. Therefore, the species of the soul should be investigated in this context.

16- Although the definition applies appropriately to the species, not all species are definable. The species of immaterial things are not known in the same way as those in speculative sciences, but some are known by intuition. Hence, the angel cannot be defined in precise terms.

17- Genus and difference can be considered in two ways: one from the real perspective, where genus and difference must be based on different natures; and another from a logical perspective, where they do not necessarily have to be different but can consider aspects of the same nature.

18- Thomas Aquinas discusses that, speaking naturally, the differences must be opposites, as matter can receive opposite forms. However, from a logical perspective, any opposition in differences is sufficient, as seen in numbers.

19- Finally, it is established that although matter does not provide the species, the nature of the form in relation to the matter must be considered. This implies that the relationship between matter and form is crucial to understanding the nature of species.

8. QUESTION 8: Whether the rational soul must be united to a body like that which man possesses

> St. Thomas presents twenty arguments by which it seems that the rational soul should not have been united to a body like the one man possesses

1- It is argued that the rational soul is a very subtle form and that the earth is the lowest body. Therefore, it would not be appropriate for such an elevated soul to be united to such an inferior body as the earthly body.

2- It is held that the human body, having reached a certain perfection, could resemble the celestial body, which is completely pure and free from contrarieties. However, if the human body resembles the celestial body, this would imply that the celestial body is more noble. Since the rational soul is superior to any form, it should be united to a celestial body, not a human one.

3- It is discussed that if the celestial body is more perfect than the rational soul, it must be related to an intelligent substance. If this substance is only a mover, the human body is more perfect in its union with the soul, as the mover merely moves without giving form. Therefore, even if there could be an intellectual substance as the form of the celestial body, a body would not be required, for the activity of the intellect does not depend on a physical organ.$_3$

4- It is asserted that every created intellectual substance has the possibility of sin, as it can turn away from the supreme good, which is God. If intellectual souls were united to celestial bodies as forms, they could sin, leading to a separation of the soul from the celestial body. Indeed, the punishment of sin is death. Consequently, this reflection would lead us to uphold the corruption of celestial bodies and the subsequent suffering of their souls in hell, something that cannot be accepted.

5- It is held that every intellectual substance is capable of attaining beatitude. If celestial bodies are animated by intellectual souls, these could also be beatified. This would imply that not only angels and humans, but also other entities, could enjoy eternal beatitude, which contradicts the teaching of the Church Fathers regarding the community of saints.

6- It is mentioned that Adam's body was suitably proportioned for the rational soul. However, the current human body is different, as it is mortal and suffers, which indicates that these bodies are not suitable for the rational soul.

7- It is said that instruments must be perfectly obedient to the mover. Since the rational soul is the noblest mover among the inferior ones, it should have a body that obeys completely. However, the human body struggles against the spirit, which is not appropriate.

8- It is suggested that the rational soul should be united to a completely spiritual body, as the human heart is the hottest of the animals by virtue of its generative capacity. Therefore, it would be more appropriate for it to be united to a completely spiritual body.

9- It is argued that the soul is incorruptible, while human bodies are corruptible. Therefore, it would not be appropriate for an incorruptible soul to be united to a corruptible body.

10- It is held that the rational soul is united to the body to constitute the human species. However, it would be more convenient for the body to which the soul is united to be incorruptible, which would avoid the need for generation to preserve the species.

11- It is asserted that for the human body to be the noblest among inferior bodies, it must resemble the celestial body, which is the most noble. But the celestial body has no contrarieties. Since human bodies do have them, they are not suitable for the rational soul.[4]

12- It is suggested that the soul is a simple form and, therefore, should be united to a simple body, such as fire or air.

13- It is mentioned that the human soul has a connection with the principles, and the ancient philosophers stated that the soul is of the nature of the principles. If the soul is not an element, then it should be united to some elemental body, such as fire or air.

14- It is held that bodies of similar parts are simpler than bodies of dissimilar parts. Therefore, the soul, being simple, should be united to a body of similar parts.

15- It is said that the soul is united to the body as form and mover. Therefore, the rational soul, being the noblest form, should be united to a very agile body, but the bodies of humans are not as agile as those of birds and other animals.

16- Plato is quoted, who says that forms are given by the giver according to the dispositions of the matter. But the human body does not seem to have the appropriate disposition for such a noble soul, as it is coarse and corruptible.

17- It is asserted that in the human soul there are intelligible forms very particular in relation to superior intelligible substances. Such forms would correspond to the operation of the celestial body, which causes the generation and corruption of these particulars. Therefore, the human soul should be united to celestial bodies.

18- It is established that nothing moves naturally when it is in its place. However, the heavens move in their place, which implies that they do not move naturally. Therefore, it is suggested that the heavens have a soul united to them.

19- It is mentioned that the act of narrating is proper to an intelligent substance. Since the heavens narrate the glory of God (Psalm 18,1), it is concluded that the heavens are intelligent and, therefore, have an intellectual soul.

20- It is argued that the soul is the most perfect form, so it should be united to a more perfect body. However, the human body seems to be the least perfect, as it lacks weapons of defense and other attributes possessed by the bodies of other animals.

> Next, St. Thomas presents an argument from authority according to which the soul should be united to a body like the human body

The creation account in the book of *Genesis* mentions that God formed man from the dust of the earth. This indicates that the human body is essentially connected to the earth and the physical world.

The assertion that man was created in God's image highlights his dignity and nobility. In Christian theology, this implies that the human being has the ability to reason, love, and relate to God, which distinguishes him from other creatures.

It is held that everything God creates is perfect and suited to its purpose. If God has created man with a rational soul that bears His image, this suggests that the union of this soul with a terrestrial body is intentional and appropriate.

It is argued that, since the rational soul is a manifestation of God's image, it is fitting for it to be united with a body that, though earthly, is part of the divine plan. This implies that there is a purpose in the creation of the human being as he is, with both his physical body and rational soul.

This objection reinforces the idea that God's creation of humanity is coherent and appropriate. The rational soul, which reflects the divine image, finds its natural place in a human body, emphasizing the importance and value of the human being in the order of creation.

> Next, St. Thomas offers his own response to the Question posed

St. Thomas presents a philosophical analysis of the relationship between the rational soul and the human body. It is established that matter exists because of form, not the other way around, which implies that the nature of the human body must be determined by the characteristics of the soul that inhabits it. Aristotle's *De Anima* is cited to affirm that the soul is not only the form of the body but also its mover and end, indicating that the rational soul, as the form of the human body, directs it and gives it purpose.

The human soul is considered the lowest in the order of intellectual substances, and unlike higher beings, it does not possess innate ideas or "intelligible species." It is described as a *tabula rasa*, which must receive knowledge of the external world through the senses. The need for the body is emphasized by arguing that the human soul needs to perceive the world through the senses, especially the sense of touch, which is fundamental for all perceptions. This sensitive capacity is essential for the operation of the intellect.

It is concluded that the human body must be designed in such a way that it is the most suitable to receive and process sensory perceptions, from which the "intelligible species" will be derived. This implies that the body must be well-balanced and have a good disposition for the sense of touch, which is essential for sensory experience. Although this sense has unique characteristics, it must be able to interact with opposites (hot and cold, wet and dry), requiring the human body to be balanced in its composition.

Furthermore, it is highlighted that nature operates in degrees, from simple elements to the human being, who is considered the most perfect result of this mix. The disposition of the human body facilitates both intellectual and sensory operations, with the structure of the brain being crucial for cognitive functions. With a larger brain relative to its size, the human being is better equipped for intellectual operations.

However, despite its perfection, the human body is subject to defects inherent in matter, which are not a failure of the creator, but natural conditions that arise from matter itself. In the creation of the human being, God granted an original justice that made the body fully subordinate to the soul while it was in communion with God. However, due to sin, the soul separates from this grace, leading the body to suffer deficiencies.

> Next, St. Thomas responds to each of the twenty arguments initially presented, which argued that the soul should not be united with a body like the human body

1- St. Thomas holds that, although the soul is the most subtle of all forms in its capacity to understand, it needs to be united to a body, which is accomplished through complexion (the combination of elements) to acquire intelligible species through the senses. This implies that the body the soul needs must have more of the heavier elements, like earth and water, to allow this union. If fire, which is highly effective in its action, did not have these elements in greater quantity, there could not be an adequate mixture.

2- The rational soul unites with a body, not because it is similar to the heavens, but because it has a similar composition. However, according to Avicenna, it unites with this type of body due to its similarity to the heavens, as he believed that lower bodies are caused by the superior ones. When lower bodies reach a composition similar to that of celestial bodies, they obtain a form similar to that of the celestial body, which is considered animated.

3- On the animation of celestial bodies, there are different opinions among philosophers and theologians. Anaxagoras held that the active intellect is completely immersed and separated, and that celestial bodies are inanimate. Other philosophers asserted that celestial bodies are animated, although some, such as Plato and Aristotle, claimed that God is something superior to the soul of the heavens. Among theologians, Origen and his followers considered celestial bodies to be animated, while others, like Damascene, maintained that they were inanimate. St. Thomas concludes that celestial bodies are moved by an intellectual substance that acts as form, having only intellectual potential, but not sensitive.

4- Although all created intellectual substances can sin according to their nature, many have been preserved by divine choice and grace so that they do not sin. This could include the souls of celestial bodies, considering that the demons who sinned are of a lower order.

5- If celestial bodies are animated, their souls would belong to the society of angels. St. Thomas mentions that he is not certain whether the sun, moon, and stars belong to the same society of angels, though some consider them to be luminous bodies, not necessarily sensitive or intellectual.

6- The body of Adam was provided for the human soul, not only according to what nature requires, but also by grace, from which humans are deprived, though nature remains unchanged.

7- The internal struggle in man, caused by opposing desires, also comes from the necessity of matter. If man has senses, he will necessarily perceive pleasures, which leads to concupiscence for those pleasures, often opposing reason. However, man was granted a remedy by grace in the state of innocence, so that the lower faculties did not oppose reason, although this gift was lost due to sin.

8- Spirits, although they are vehicles of virtues, cannot be organs of the senses. Therefore, the human body could not be composed solely of spirits.

9- Corruptibility comes from the defects inherent in the human body due to the nature of matter, especially after sin, which deprived humanity of the assistance of grace.

10- It is better to consider what is most suitable for the end, rather than what arises from the necessity of matter. It would be better if the body of an animal were incorruptible, if the matter composing it allowed for such a condition.

11- Bodies that are closer to the elements and have more opposition, such as stones and metals, are more durable, because their harmony is lesser, which makes decomposition more difficult. However, in animals, the duration of life depends on the humidity not drying out easily and the heat not being extinguished, as life resides in heat and moisture. This is found in humans according to the balance required by their composition.

12- The human body cannot be a simple body, nor can the celestial body be one due to the capacity of sensory organs, especially the sense of touch. Nor can it be a simple elemental body, as the element contains opposites in act. Therefore, the human body must be a balanced body.

13- The ancient philosophers believed that the soul, which knows all, must be similar to all things. That is why they thought it was part of the nature of the elements, which they considered the principle of all things, so that the soul could know everything. However, Aristotle demonstrated that the soul knows everything inasmuch as it is similar to all things in potential, not in act. Therefore, the body to which the soul is united should not be at an extreme, but in the middle, so that it can be potentially opposed.

14- Although the soul is simple in essence, it is multiple in power, and the more perfect it is, the more so. Hence, it requires an organic body that has different parts.

15- The soul is not united to the body by local movement; rather, the local movement of man and other animals is directed toward the preservation of the body united to the soul. The soul is united to the body to understand, which is its main operation, and therefore it requires that the body be well-disposed to serve the soul in what is necessary for understanding, having agility and other qualities that its disposition allows.

16- Plato affirmed that the forms of things subsist by themselves, and that the participation of forms in matter is for perfecting them, not for the forms that subsist by themselves. For this reason, he concluded that forms are given to matters according to their merits. However, according to Aristotle, natural forms do not subsist by themselves; therefore, the union of form with matter is not by matter, but by form. It is not a matter of matter being disposed to receive the form; rather, matter must be disposed so that the form may be such. Thus, the human body is disposed according to what corresponds to such a form.

17- Although the celestial body is the cause of the particular that is generated and corrupted, it is, however, the cause of the general. For this reason, determined agents are required for determined species. Thus, the motor of the celestial body does not need to have particular forms, but universal ones, whether it be a soul or a separate motor. However, Avicenna considered that the soul of the sky should have imagination in order to understand the particular. Since it is the cause of the movement of the heavens, it must know the "here" and the "now." But this is not necessary, first, because the movement of the heavens is uniform and does not present obstacles; so a universal conception is sufficient to cause such movement. Particular conception is necessary in the movements of animals due to the irregularity of their movement and the obstacles that may present themselves. Second, superior intellectual substances can comprehend the particular without the need for a sentient power.

18- The movement of the heavens is natural due to the passive or receptive principle of movement, as such a body naturally corresponds to such movement; but the active principle of this movement is some intellectual substance. What is said about no body moving naturally in its existing place is understood in relation to a body moving in a straight line, that is, changing place not only in reason but also in the subject. A body that moves in circles does not completely change place in the subject.

19- The human body must be a composite body, not an elemental body, because its organic structure and senses require a specific organization.

Therefore, the perfection of the human body lies in its complexity and its ability to act as a whole, which allows the soul access to the various sensory experiences necessary for knowledge.

20- The ancient philosophers considered that the perfection of both the body and the soul should be in harmony, as each depends on the other to fulfill its purpose. The human body, with its complexity and organization, allows the soul to perform its function of knowledge. Therefore, the essence of the human body must be configured in such a way that it provides a conducive environment for the development and activity of the soul.

9. QUESTION 9: Whether the soul is united to corporeal matter through a medium

> St. Thomas Aquinas presents nineteen arguments suggesting that the soul seems to be united to corporeal matter through a medium

1- The soul appears to unite with the body through powers or faculties, according to *De spiritu et anima*.₅ These powers are distinct from the soul's essence, implying that the soul unites with the body through something other than its essence.

2- It is argued that the soul unites with the body through its powers as a mover (agent of movement) but not as a form. However, this distinction is contested. It is asserted that the soul is both the form of the body (giving existence) and its mover (initiator of action) at the same time. Therefore, it is inappropriate to separate the soul's function as mover and form.

3- If the soul were united to the body as a mover only accidentally, it would not form a true unity with the body. But the soul unites with the body in itself, without an intermediary, emphasizing that this union is essential and not accidental. Thus, the soul is a mover in that it is united to the body without a medium.

4- The operations of the soul pertain to the composite, not to the soul alone. In this sense, there is no intermediary between the soul and the body for performing these operations.

5- As the form of the body, the soul does not unite with just any matter, but with matter suitably disposed. This disposition is provided through its own accidents, such as heat and dryness in fire. The soul's powers, being its proper accidents, could serve as a medium for its union with the body.

6- An animal moves itself and is divided into two parts: that which moves and that which is moved. The soul is the moving part, and the body,

THE HUMAN SOUL

being matter and form, is the moved part. For this reason, it appears that the soul is united to corporeal matter through some intermediate form.

7- The definition of any form includes its proper matter. Thus, in the soul's definition, the physical and organic body is included as its proper matter, which suggests that the soul unites with the body through a form that perfects the matter first.

8- In *Genesis*, it is stated that God formed man from the earth and breathed into him the "breath of life." This suggests a prior form in matter that precedes the soul's union, acting as an intermediary.

9- Forms unite with matter according to the potential capacity of the latter. First, matter is in potency for elementary forms before other forms. Thus, the union of the soul would require elementary forms as intermediaries.

10- The human body is a mixture of elements, and these forms must remain in the mixture; otherwise, the elements would disintegrate. Therefore, the soul would unite with matter through these elemental forms.

11- The intellectual soul unites with the body as a form in that it is intellectual, and the act of understanding involves mediating powers. This suggests that the soul unites with the body as a form through its powers.

12- The soul does not unite with just any body, but with a proportionate one. This proportion is the medium through which the soul unites with the body.

13- The soul acts in the body primarily through the heart, which is closest to the soul regarding its powers. This suggests that the soul could unite with the body through the heart as an intermediary.

14- Given that the body has diverse parts and the soul is simple, it would seem logical that the soul unites first with a specific part of the body and then with the rest through this part.

15- The soul, being superior to the body, uses higher powers (such as the intellect) which are connected to the body by lower powers (such as imagination and sense). This suggests that the body unites with the soul through simple elements like spirit and humors.

16- If removing something breaks the union between the soul and the body, this indicates an intermediary. When the "vital spirit" or natural heat disappears, the body and soul separate, suggesting these elements are the medium of union.

17- The soul unites with a specific body that has determined dimensions, and these dimensions could act as intermediaries in the union.

18- Since the soul and body are very different in nature (one being incorporeal and simple, the other corporeal and composite), a mediation seems necessary to unite them.

19- The human soul is similar in nature to separated intellectual substances that move celestial bodies. Thus, it is suggested that the human body possesses something of the nature of celestial bodies that facilitates the soul's union.

> Next, St. Thomas presents an argument from authority that the soul does not unite with the body through an intermediary

According to Aristotle, the form unites directly with matter, so being the form of the human body, the soul unites with it immediately.

> St. Thomas then offers his own answer to the Question posed

St. Thomas's answer is a profound analysis of the relationship between form and matter, particularly in the context of the essence of beings and their existence. In this text, St. Thomas argues that "being" is the most intimate and essential aspect of things, with the form conferring actual existence on matter.

First, St. Thomas argues that substantial form is what grants matter its being directly and essentially. The substantial form determines what something is; without it, matter would lack a concrete being. For example, in the case of a human being, the soul is the form that grants matter (the body) its specific nature of "human." In this sense, substantial form differs from accidental forms, which only give matter additional characteristics (such as color or size) but not its essence.

St. Thomas also rejects the idea of intermediate substantial forms between matter and substantial form. This critique addresses certain philosophical positions proposing a hierarchy of forms, where an initial form could give matter existence at a lower level, and other forms would add further perfections. According to him, there can only be one substantial form that grants being to matter in its entirety; that form is what makes matter a specific "something."

Moreover, it emphasizes that the forms of natural beings can be understood as a kind of hierarchy, where one form may be more perfect than another. This perfection is understood in terms of how forms constitute matter in different degrees of being, from mere corporeal existence to animated and rational existence. In this sense, matter, when it exists in a corporeal state, is a kind of foundation that can be perfected by incorporating higher forms, such as the soul.

Saint Thomas also addresses how the substantial form not only determines the existence of matter but also acts as a principle of operation. The form grants matter not only being but also operational capacities, and this varies according to the degree of perfection of the form. More perfect forms have a greater capacity for action, which results in a diversity of functions in living beings. For example, in humans, different parts of the body (such as the heart or lungs) have specific functions that depend on the unity that the form (the soul) bestows upon them.

Finally, Saint Thomas defends the idea that the soul and body are united in such a way that the soul acts as the form of the body without intermediaries; that is, the soul gives life and specificity to the body directly. This opposes theories that propose the existence of intermediate elements that connect the soul and body, such as certain humors or spirits.

> Following this, Saint Thomas responds to each of the nineteen initial arguments, which considered that the soul was united to the body by some intermediary

1- In the first argument, Saint Thomas explains that the faculties of the soul are qualities that allow it to act and are thus positioned between the soul and the body in terms of movement. However, they do not mediate in terms of the function of giving existence. He also clarifies that the book attributed to Augustine, *De spiritu et anima*, is not actually by him, and that the author of that text held that the soul is its own power, so the objection presented does not apply.

2- Regarding the second argument, Saint Thomas clarifies that, although the soul is form in terms of act and mover, its effects differ according to the function it fulfills. As form, it has one effect, and as mover, another; thus, the distinction is valid.

3- In the third objection, he explains that between a mover and what it moves, there is no essential unity; however, in the case of the soul and the body, there is, as the soul is the form of the body.

4- In the fourth response, Saint Thomas clarifies that with regard to the joint operation of the soul and body, there is nothing between the soul and the parts of the body. There is only a part of the body through which the soul acts first, and other parts participate in that operation.

5- In the fifth objection, it is argued that the accidental dispositions necessary to prepare matter for a form are not completely intermediate between form and matter. They serve as mediators between the form as the final perfection and matter at a lower degree of perfection. The soul's powers are accidental and proper to it, so they do not function as dispositions toward the soul but as relations between different levels of power.

6- In the sixth argument, Saint Thomas acknowledges that the soul and body are divided into the moving and the moved part, which is true, but the soul moves the body through perception and desire. However, intellectual perception in humans only moves the body indirectly through sensory perception.

7- The seventh point addresses the relationship between the physical organic body and the soul, establishing that the body acts as matter for the soul, not through an intermediate form, but because the body derives this from the soul itself.

8- In the eighth response, it is clarified that, in the creation of man in *Genesis*, the sequence of "forming man from the dust" and then "breathing the breath of life" does not imply a temporal separation but a natural order.

9- The ninth explanation states that matter is potentially ordered toward forms without this meaning that it receives multiple substantial forms sequentially; rather, each higher form requires the qualities of a lower form to manifest.

10- In the tenth objection, Saint Thomas discusses the position that elemental forms would be actively present in the composite. However, he explains that the elemental forms remain by virtue of their accidents, not in their essence, in the composite.

11- The eleventh response addresses that although the soul is the form of the body, its intellectual operation does not bind it completely to its material nature.

12- In the twelfth argument, it is established that the proportion between soul and body resides in the proportional elements themselves and does not require an intermediary between them.

13- In response to the thirteenth argument, Saint Thomas affirms that the heart acts as the primary instrument through which the soul moves the body, and although the soul is united to all parts of the body, it is particularly united to the heart as mover.

14- In the fourteenth argument, it is acknowledged that the soul, being a simple form, possesses multiple faculties that allow for various operations. This diversity affects the perfection of each part of the body according to its respective functions.

15- In the fifteenth, he explains how the lower faculties of the soul help the higher ones to operate, just as the body connects with the soul through the higher functions of the body.

16- The sixteenth response holds that the union between the soul and body dissolves when natural dispositions, like heat and moisture, are altered; these dispositions act as intermediaries between the soul and body.

17- In the seventeenth argument, it is clarified that dimensions only apply to matter once it is constituted with the substantial form, which, in the case of humans, is the soul.

18- In the eighteenth, Saint Thomas refutes the notion that soul and body are separated as entities of different genera or species; the soul is the form that gives being to the body and unites directly with it.

19- Finally, in the nineteenth argument, it is established that the human body shares characteristics with the celestial body regarding its balance of qualities, although there is no intermediary entity between the soul and body.

10. QUESTION 10: Whether the soul exists in the entire body and in each of its parts

> Saint Thomas presents eighteen arguments suggesting that the soul does not reside throughout the body and within each of its parts

1- The soul functions as the perfection of the body, but only of the "organic" body (structured for life), since, according to Aristotle, the soul is the act of a physical body that has life potentially. As not all parts of the body are "organic," it seems the soul might not be in every part.[6]

2- The soul's form is simple, while the body, with its various parts, is multiple and diverse. Since form must correspond with matter, a simple essence could not align with matter as diverse as the body, suggesting that the soul may not be present in all parts.

3- If the soul is entirely in each part of the body, then there would be "nothing" of the soul outside that part, implying it could not simultaneously reside in all parts.

4- Aristotle compares the soul to a well-ordered political system. Just as in a city the ruler need not be in every part, so too in the body, the soul resides in a central principle, and each part fulfills its function without the need for the soul to be in each one.

5- Aristotle, in *Physics*, argues that the "mover" of the heavens must be at the circumference and not at the center, because the motion is faster the farther one is from the center. Applying this idea to the body, it is stated that the soul must reside in the heart, as it is the part of the body where movement is most apparent. Therefore, the soul is located in the heart of the animal.

6- Aristotle mentions that plants have a vital principle between the superior and inferior parts. By analogy, the soul should occupy a central location in the body, such as the heart, rather than every part.

7- When a form is present in a whole and each part, as with fire, each part also takes the form's designation (e.g., every part of fire is fire). But we cannot say each part of the body is "animal." Therefore, the soul does not seem to be in every part of the body.

8- The faculty of understanding belongs to the soul, but understanding is not located in any specific part of the body. Thus, it appears the soul is not wholly present in each part.

9- Aristotle proposes a correspondence between soul and body, suggesting that each part of the soul should correspond to a part of the body. Thus, if the soul is in the whole body, it may not be entirely in each part.

10- One might object that Aristotle speaks of the soul's parts as motor rather than form. However, Aristotle himself states that if an eye were an animal, sight would be its soul. Therefore, a part of the soul should reside in the body, not solely as a mover but as form.

11- The soul is the life principle in the body. If it were in all parts, each part would receive life directly from the soul without reliance on others, yet we know body parts depend on the heart for life.

12- The body's movement is attributed to the soul accidentally, and the body can move and rest in different parts. If the soul were in every part of the body, it would move and rest simultaneously, which is impossible.

13- The soul's powers are rooted in its essence, and if it were in every part of the body, then all powers would be in each part, which is not the case, as, for instance, hearing resides only in the ear.

14- Everything contained within something else is there according to the mode of that container. If the soul is in the body, it should exist in it according to the body's mode, which does not allow one part to be in another; hence, the soul may not be in each part of the body.

15- Some segmented animals, like worms, can live when cut into parts because the soul persists in each section. However, humans and higher animals do not survive division, indicating that the soul is not in every part of their bodies.

16- Just as the form of a house is not in each brick but in the whole, the soul, as the body's form, should reside in the whole and not in each part.

17- The soul grants existence to the body through its simple essence. As from one, only one can proceed, if the soul were in each part, it would confer uniform existence to all parts, which does not seem to be the case.

18- The union between form and matter is more intimate than that of an object with the space it occupies. But just as a spiritual being cannot be in several places at once, the soul should not be in different parts of the body simultaneously.

> Next, Saint Thomas presents arguments from authority affirming that the soul is present throughout the body and within each of its parts

Saint Augustine, in his work *De Trinitate*, holds that the soul is fully present in the entire body as well as in each of its parts. This implies that the soul's essence is entirely present in each part of the body, contradicting the idea that it cannot be in all parts.

The soul does not grant existence to the body except insofar as it unites with it. Since the soul confers being to both the entire body and each of its parts, it follows that the soul must be present in both the complete body and each part.

The soul can only act where it is present. Since the soul's operations manifest in each part of the body (such as in the movement of limbs or the perception of senses), it is concluded that the soul must be present in each of these parts.

> Saint Thomas then offers his own answer to the Question posed

The text examines the relationship between soul and body from a philosophical perspective, emphasizing that the soul, as the body's form, does not unite with it through any particular part but is present as a whole within every part. This relationship implies that the soul is responsible for conferring existence and specificity to each component of the body. It is argued that the body, understood as a "natural whole," derives its unity from a single form that perfects it, as opposed to inorganic constructions, where unity arises from the mere aggregation of parts.

Additionally, it is argued that each part of the body receives its being and form from the soul. When the soul separates from the body, none of its parts can continue to exist in the same way. This relates to a critique of Platonic ideas suggesting that sensible beings obtain their existence through participation in separate forms. In this context, the Question is addressed of how totality can be attributed to a form, identifying three different modes of conceiving totality: quantitative division, referring to size or quantity; essential division, where form and matter are parts that constitute a composite; and division by power or virtue, considering how the operations of a form may differ according to the parts involved.[7]

It is concluded that the soul, being the form of the body, is present totally and perfectly in each part of the body in terms of its essence and specific perfection. However, a distinction is made regarding its operative power, as the soul does not act fully in each part. For example, functions such as understanding and will do not depend on specific bodily organs, suggesting that the soul possesses a capacity that transcends physical interaction with the body. In this sense, the text offers a comprehensive view of the relationship between the soul and the body, highlighting the need for both for the full realization of the human being.

> Next, Saint Thomas responds to each of the eighteen arguments which suggest that the soul is not present in the entire body and in each of its parts

1- Saint Thomas explains that matter exists for the form, and the form is oriented towards a particular operation. Therefore, the matter must be adequate for the operation of the form. Just as the matter of a saw must be iron for its hardness, the soul, given its perfection, requires a body composed of parts suitable for its various operations. The whole body is considered as an organ, and the parts exist for the whole. Thus, it cannot be said that each part of the body is an organism by itself, but each part has a relationship with the whole body and with the operations of the soul.

2- Here it is clarified that, since matter exists for the form, this form grants being and species to matter in relation to its operation. Although the body is one and simple in essence, it presents diverse parts which are perfected by the soul in different ways, according to their functions.

3- Saint Thomas responds that the soul is in one part of the body in a particular way, but this does not mean that there is nothing of the soul outside that part. What is affirmed is that the soul is not outside the body as a whole, because the soul perfects the whole body.

4- It is mentioned that Aristotle speaks of the soul in relation to its capacity for motion, and that the principle of motion is found in a specific part of the body, namely, in the heart. Through this part, the soul moves the entire body.

5- Saint Thomas clarifies that the motor of the heavens is not limited by its location in its essence. Aristotle seeks to indicate where it is in relation to the principle of movement. Therefore, in terms of the cause of movement, the soul is in the heart.

6- At this point, it is argued that, in plants, the soul is said to be in the middle of what is upwards and downwards, acting as the principle of certain operations. This statement is similar in animals.

7- Here, it is distinguished that not every part of an animal is a complete animal, unlike fire, where each part retains all the operations of fire. In animals, especially the more perfect ones, not all operations are performed in every part.

8- Saint Thomas confirms that the reasoning presented shows that the soul is not present in every part of the body in terms of its complete capacity, which is true.

9- The parts of the soul are understood in terms of their power, not their essence. Thus, just as the soul is in the whole body, a part of the soul is in a part of the body, with each organ corresponding to a specific operation.

10- Saint Thomas indicates that the power of the soul is rooted in its essence. Therefore, where there is the power of the soul, there also is its essence. Aristotle's statement about the eye is understood in the context of the power of the soul, not separated from its essence.

11- When the soul operates in other parts of the body through a single power, the body must be disposed so that it corresponds to the being of the soul through its action. This means that the disposition of the other parts depends on one main part, the heart, from which the life of the rest of the body emanates.

12- The soul does not move or rest, independently of the movement or rest of the body, except by accident. This is not problematic, as something can move and rest simultaneously by accident, like an object moving in a ship.

13- Although all the powers of the soul have their root in its essence, each part of the body receives the soul according to its own mode. Therefore, the soul presents itself in different ways in the various parts of the body, without the need to be present in every part with all its powers.

14- When it is said that something is in another according to the mode of its capacity, it does not refer to the nature of what is in it, but to the capacity to receive. Thus, the water does not have the nature of the jar in which it is, just as the soul does not need to have the nature of the body in which it resides.

15- It is clarified that annelid animals can live even if a part is cut off, not only because the soul is in each part of the body, but because their soul is imperfect and has fewer functions, requiring less diversity in the parts, allowing the soul to remain in a cut part. This is different in more perfect animals.

16- It is pointed out that the form of a house is accidental, it does not provide being and species to each part as the soul does, which is a substantial form of the body, granting being and species to the whole and to the parts.

17- Saint Thomas clarifies that although the soul is one and simple in essence, it has the capacity to perform diverse operations. This diversity in the parts of the body responds to the different functions the soul must fulfill.

18- Finally, it is distinguished that the simplicity of the soul and the angel should not be considered as the simplicity of a point in a continuum, since what is simple cannot occupy multiple places in the continuum simultaneously. Both the angel and the soul are considered simple in that they lack quantity, and their relationship with the continuum is through the contact of their powers, allowing the soul to be present in each part of its perfect being.

11. QUESTION 11: Whether the rational, sensitive, and vegetative souls in man are substantially one and the same

> Saint Thomas presents twenty arguments suggesting that the soul in man is not one and substantially the same

1- The act of the soul determines the existence of the soul itself. In the embryo, the activity of the vegetative soul precedes that of the sensitive soul, and the latter precedes that of the rational soul. This suggests that, in terms of substance, the vegetative soul is prior and distinct from the sensitive soul, and the sensitive soul is distinct from the rational soul.

2- It could be argued that the activities of the vegetative and sensitive souls in the embryo come from an external force, that of the parents' souls. However, this is rejected because a finite agent (such as a parent) cannot influence something at an indefinite distance; the movements and operations observed in the embryo must come from an internal principle, not the parents' force.

3- Aristotle affirms that the embryo is first an animal before being considered human. Since an animal is defined by having a sensitive soul, this implies that the sensitive soul is present in the embryo before the arrival of the rational soul.

4- To live and to feel are functions that require an internal principle (the soul). Since the embryo has life and sensations before having a rational soul, these activities cannot be attributed to an external soul (that of the parents), but to a soul that is already present in the embryo.

5- Aristotle teaches that the soul is the cause of the living body not only as form, but also as efficient and final cause. If this is true, then the soul must be present in the body in formation before the infusion of the rational soul. Therefore, there must be a soul in the embryo before the infusion of the rational soul.

6- The formation of the body must be attributed to the soul that is in the embryo, not to that of the father. Since a living body moves by itself and its generation is a type of movement, the principle that forms it must be the soul within the embryo.

7- The embryo grows, which is a type of movement. Since living beings move by themselves, this implies that there must be an internal principle (the soul) responsible for this growth, and it does not come from an external influence.

8- Aristotle notes that in the embryo there is a soul that is first nutritive and then sensitive. This implies that there is already a form of soul present, which strengthens the idea that the embryo has an internal principle.

9- One might argue that Aristotle refers to the soul in the embryo as being potential, not actual. However, only entities in act can act. Since the embryo shows actions of the soul, there must be an actual soul present.

10- There can be no contradiction between what comes from external and internal sources. If the rational soul is considered external to man, while the vegetative and sensitive souls are internal (coming from the semen), then they cannot be the same substance.

11- It is not possible for what is substantial in one being to be merely accidental in another. If the sensitive soul is substantial in brute animals, it cannot be merely potential in humans, as powers are accidental properties of the soul.

12- Since the human being is a nobler animal than the brute animals, and since he is defined as such by having a sensitive soul, it is logical to conclude that the sensitive soul in man is a substance in itself, not just a potential principle.

13- There can be no substantial identity between the corruptible and the incorruptible. While the rational soul is incorruptible, the sensitive and vegetative souls are corruptible, indicating that they cannot be the same substance.

14- One could claim that the sensitive soul in man is incorruptible. However, this would imply that it is of a different nature than the sensitive soul in other animals. Therefore, if they are of different kinds, they cannot belong to the same species.

15- It is not possible for what is rational and what is irrational to be the same substance, as this would be contradictory. Since the sensitive and vegetative souls are irrational, they cannot be the same as the rational soul.

16- The body is related to the soul. However, there are various functions within the body that require different principles of operation. This suggests that there cannot be just one soul in the human being.

17- The powers of the soul emanate from its essence. However, from a single essence, different powers cannot derive; if there were only one soul in the human being, there could not be powers associated with organs and powers not associated.

18- The definition of a species is based on matter and form. The genus of man is "animal" and his difference is "rational." If man is defined by the sensitive soul, then the sensitive soul must be compared to the rational as matter is compared to form, implying that they are not the same substance.

19- Both man and the horse belong to the genus "animal," defined by the sensitive soul. However, if the sensitive soul in horses is not rational, then neither is it in man.

20- If the rational, sensitive, and vegetative souls are identical in substance, then wherever one of them is, all must be present. However, this is false; for example, in the bones, there is only a nutritive (vegetative) principle, not a sensitive one, which proves they are not identical in substance.

> Next, Saint Thomas presents an argument from authority according to which in man, the soul is one and the same rational, sensitive, and vegetative substance

This argument against the idea of two souls in a human being is based on a statement by the priest Genadius (second half of the 5th century) in his work *De ecclesiasticis dogmatibus*, according to which it is not possible to have two souls in a single human being, as some (mentioning Iacobus and other Syrians) have suggested. Instead of accepting this idea of two souls—one that animates the body and another that relates to reason—the argument defends the notion that there is only one soul in the human being.

The explanation of the argument is as follows:

1- Unity of the soul: It is asserted that a human being cannot have more than one soul. The central claim is that there is a single soul that enlivens the body and, at the same time, organizes and directs its functions through reason.

2- Functions of the soul: This argument emphasizes that the soul is not only responsible for giving life to the body but also acts as the principle that orders and directs the rational capacities of the human being. In other words, it is argued that the human soul has a dual function: on one hand, it acts as the vital force, and on the other, it deals with intellectual and rational capacities.

3- Refutation of plurality: By asserting that there is only one soul, the argument refutes the idea that two souls could coexist in a single human body (one to animate and another to reason). By insisting on the unity of the soul, it is suggested that all the capacities and functions of the human being (vegetative, sensitive, and rational) arise from this one soul.

4- Philosophical implications: The deeper implication of this argument is that the distinction between the vegetative, sensitive, and rational souls does not mean that there are multiple souls; rather, it means that a single soul can manifest different powers and functions in different contexts.

Saint Thomas' response to the Question

Saint Thomas acknowledges that there are different positions on the nature of the soul, both among contemporary thinkers and ancient philosophers. He mentions Plato, who proposed that there are multiple souls in the human body. According to Plato, the soul joins the body as a motor that moves it, not as a form that defines it. He compares the relationship between the soul and the body to that of a sailor and a ship, where different motors (souls) are required for different actions, but this does not contradict the unity of the ship.

Although Plato seems to allow for multiple souls in one body, Saint Thomas argues that this would imply that the human being is not a simple unity, contradicting the notion that an individual is a singular entity. If the soul only acts as a motor, then the union between the soul and the body would not be essential, and therefore there would be no true generation or corruption when a body gains or loses a soul. Santo Tomás concludes that the soul must be united to the body not only as a motor but also as a form, meaning that the soul is essential to the nature of the human being.

Even accepting that the soul is a form, Plato's followers would maintain that multiple souls can exist in a human or animal, as they posit that universals (like the idea of "animal") are separate forms. This idea suggests that there is a form (soul) for each type of being. For example, Socrates would be an animal by one form and a human by another, leading to the conclusion that sensitive and rational souls are substantially different.

Saint Thomas refutes this by asserting that unity cannot come from multiple substantial forms. If something is defined by different forms, then statements about its nature are merely accidental, which denies the true identity of the being. If it is said that something is both "man" and "animal" under different forms, this implies that one form accidentally predicates the other. This creates confusion about the true nature of what is being described.

For a being to be considered a unity, there must be a principle that unites the various forms; otherwise, it would become a mere collection, like a pile of objects that are many in one but are not a single being. The argument holds that if the sensitive soul of an individual defines him as "animal," then it must be a substantial form. This implies that this soul must give true existence to its body, not just in a relative sense.

If the rational soul were different in essence, it could not give existence or being to the body, but would only give it relative existence, which would make it an accidental form rather than a substantial one. Santo Tomás concludes that in the human being, there must be only one substantial soul that is rational, sensitive, and vegetative. The reason is that the rational form integrates and perfects the other forms, giving matter what it needs to be a complete being.

Finally, he notes that when one power of the soul is intensified, it may interfere with the operation of another, suggesting that all the powers must be rooted in a single essence of the soul, confirming the unity of the soul in the human being.

Saint Thomas' response to each of the twenty arguments

1- Saint Thomas addresses the idea that before the existence of the rational soul in the embryo, it only has a "formative power" coming from the soul of the parents. He clarifies that while this "power" may be responsible for certain operations in the embryo, it does not explain all its functions. In reality, the embryo shows not only bodily formation but also capacities like growth and perception, which belong to the soul. He proposes that this "power" can be considered a part of the developing soul, but it cannot be seen as a complete soul, since the embryo is not a complete human being. Therefore, it is maintained that the embryo must undergo development that includes several stages of generation, each with a form that is perfected, from a vegetative soul to a rational one.

2- This objection discusses the nature of the "virtue" in the father's semen, which acts intrinsically and not extrinsically. Saint Thomas maintains that, unlike an external force that can only affect within certain limits, the "virtue" of the semen can generate life regardless of distance. He emphasizes that, although it has been argued that the mother is not the active principle, the influence of the semen is what allows the development of the embryo.

3- The "virtue" present in the semen is equated with the essence of the soul, thus allowing the embryo to be considered an animal. This shows that, although the embryo is in a developmental stage, it has the capacity to be recognized as a living being.

4 to 8- Saint Thomas indicates that the responses to objections 4 to 8 are similar and based on the same logic: it is acknowledged that the soul of the embryo, although imperfect, acts in such a way that the development of basic functions can be observed.

9- Here it is stated that, although the soul is present in the embryo, it does so in an imperfect way, which is also reflected in its operations. This means that the capacities of the embryo are limited in comparison to a developed human being.

10- The nature of the soul in man, which encompasses both the vegetative, sensitive, and rational aspects, is of an external origin. This contrasts with animals, whose sensitive soul is intrinsic and defined according to their nature.

11- It is clarified that the sensitive soul in human beings is not an accident, but a substance, as it shares the same essence with the rational soul. However, the sensitive capacities are accidental, meaning they may vary or not be essential to the identity of the being.

12- It is argued that the sensitive soul in human beings has a higher dignity compared to animals, since in man, sensitivity is accompanied by rationality, which is not the case with other living beings.

13- Saint Thomas asserts that the sensitive soul in human beings is incorruptible because its substance is that of the rational soul. Although some may think that sensitive capacities are corruptible, the essence of the soul remains.

14- When comparing the sensitive soul of humans and animals, they are not of the same type in terms of genus or species, unless it is a logical comparison in some abstract sense. Sensitive souls are composite and, therefore, corruptible.

15- It is distinguished that the sensitive soul in human beings is not irrational; rather, it is simultaneously sensitive and rational. While some of its capacities may seem irrational, they serve the reason.

16- Although there are multiple organs in the body performing various functions of the soul, they all depend on the heart as their principal organ, reinforcing the idea of a unity in the essence of the soul.

17- The powers that emanate from the human soul manifest through the organs, but there are also capacities that go beyond the corporeal, indicating the dual nature of the soul.

18- It is argued that a single form can give different degrees of perfection to matter. As the body develops, it remains material until it reaches perfection in the sensitive being, showing that the essence of the animal is derived from matter, while rationality comes from form.

19- It is clarified that, in the classification of beings, the animal as such is neither rational nor irrational, but the human being (rational animal) is different from irrational animals.

20- Although the operations of the sensitive and vegetative soul are different, it is not required for them to manifest simultaneously. The different functions can be expressed through different parts of the body, such as sight through the eyes or hearing through the ears.

12. QUESTION 12: Whether the soul is its powers

> St. Thomas presents seventeen arguments according to which it seems that the soul is its powers

1- The first argument, taken from the book *De spiritu et anima* (see Note 5), establishes that the powers of the soul are identical to the soul itself, since the virtues of the soul (such as prudence and justice) are not accidental, but are considered essential parts of its being. This implies that the soul is its powers.

2- The second argument points out that the soul receives different names according to its functions (vegetative, sensitive, rational, etc.), but there is no change in its essence. This reinforces the idea that, despite its different actions, the soul remains the same in all its powers.

3- The third argument states that the three capacities (memory, intellect, and will) are the soul itself, and this notion extends to the other powers. This implies that the soul cannot be separated from its powers, as they are constitutive of its being.

4- The fourth argument cites St. Augustine, who holds that memory, intellect, and will are one essence of the soul. This implies that the powers are not accidental, but fundamental to the nature of the soul.

5- The fifth argument maintains that, since the powers can act not only on the soul but also on other objects, they cannot be considered mere accidents. Therefore, they must be seen as essential parts of the soul.

6- The sixth argument links the image of the Trinity with the powers of the soul, suggesting that these are intrinsic to the nature of the soul. This implies that the powers are essential to understanding the essence of the soul.

7- In the seventh argument, it is stated that the powers of the soul are necessary and intrinsic, as they cannot be absent. This reinforces the idea that they are an essential part of the soul, not merely accidents.

8- The eighth argument establishes that the powers of the soul are principles of substantial differences. This implies that they are not mere accidents, but are fundamental to defining the nature of the soul and its relation to the body.

9- In the ninth argument, it is stated that the substantial form, which is the soul, acts through its powers. This implies that the powers are not different from the soul itself, but are its way of acting and existing.

10- The tenth argument says that the principle of being and operating is the same in the soul, suggesting that its essence is the foundation of its powers. This implies that the soul is its power, as it is the principle of its actions.

11- In the eleventh argument, it is mentioned that the soul is both possible intellect and agent, and that its existence in potency and act refers to the same reality of the soul. This reinforces the idea that the soul is inseparable from its powers.

12- The twelfth argument compares the soul to prime matter, indicating that, just as prime matter is potential for forms, the soul is potential for intellectual realities. This implies that the soul is its powers, as they are what allow it to realize its nature.

13- In the thirteenth argument, it is stated that the human being is intellect by virtue of his rational soul, which implies that the soul and its power to understand are one and the same thing.

14- The fourteenth argument maintains that the soul is the first act of its operations, suggesting that it acts through its powers. This implies that the soul cannot be separated from its powers, as these are the forms through which it acts.

15- The fifteenth argument establishes that the powers are consubstantial parts of the soul. This implies that they are essential to the constitution of the soul and not mere accidents.

16- The sixteenth argument points out that a simple form, like the soul, cannot be the subject of accidents. This implies that the powers of the soul are intrinsic and not accidental.

17- In the seventeenth argument, it is proposed that if the powers were accidents of the soul, they should derive from its essence. However, since the soul is simple, it cannot be the cause of the diversity observed in its powers. This reinforces the conclusion that the soul is in itself its powers.

> Next, St. Thomas presents two arguments from authority according to which the soul is not its powerS

1- An analogy is established between essence and being, and between power *(potentia)* and action. It is argued that, just as being and acting are interdependent in their relationship, so too are power and essence. The conclusion drawn is that, if in God essence and being are identical, then essence and power must also be identical in God. Therefore, it is concluded that the soul is not its power, since the identification of power and essence is found only in God.

2- It is stated that no quality is substance, and it is pointed out that natural power is a type of quality. This is based on the classification of qualities in Aristotle's categories. Since natural powers are qualities and not substances, it is concluded that the powers of the soul are not its essence, as they cannot be considered part of its constitutive being. This reinforces the idea that the soul is not identical to its powers, but they are distinct from its essence.

> Next, St. Thomas offers his own response to the proposed Question

St. Thomas begins by noting that there are various opinions about whether the soul is its own power. Some argue that the soul is indeed its power, while others maintain that the powers of the soul are merely properties of it. This introduction establishes that there is a debate about the relationship between the soul and its powers.

He then defines what power is, stating that it is nothing more than a principle of operation, either of action or passion. Here, St. Thomas distinguishes between the principle that acts and the one that receives the action, emphasizing that power is not the subject that performs the action, but the principle through which the action is performed. He uses examples such as "constructive craftsmanship" in the builder and "heat" in fire to illustrate how power manifests in action.

Next, he addresses the perspective of those who claim the soul is its power, explaining that it is understood that the essence of the soul is the immediate principle of all its operations. Therefore, it is maintained that the human being acts (understands, knows, feels, etc.) through his essence. Each type of action is named according to the operation it performs, such as sense in relation to feeling and intellect in relation to understanding.

However, Saint Thomas refutes this opinion, pointing out that everything that acts does so according to its current reality. In other words, an agent acts according to what it is at that moment. He uses fire as an example, which heats not because of its light, but because of its heat. This implies that the principle of action must correspond to the nature of the agent, so when acting, the agent must be in conformity with its essence.

He explains that since what acts does not refer to the substantial essence of the thing, it cannot be that the principle of action is part of the essence of the thing. This is made evident in natural agents, which operate through the transformation of matter into a form, implying that action occurs through an accidental principle that corresponds to the disposition of the matter.

Saint Thomas goes on to clarify that the accidental form acts by virtue of the substantial form, like an instrument; otherwise, it could not induce a substantial form. In nature, no principles of action are observed without active and passive qualities that operate through substantial forms, suggesting that actions are directed not only toward accidental dispositions but also toward substantial forms.

He then mentions that if there were an agent that could substantially produce something directly (like God, who creates substances), such an agent would act by its own essence, and in this case, there would be no distinction between active potency and essence.

Regarding passive potency, he points out that passive potency directed toward a substantial act belongs to the genus of substance, while that directed toward an accidental act belongs to the genus of accident in a reduced manner. This implies that there are distinctions in how potency is classified based on its relation to the corresponding act.

Next, he maintains that the powers of the soul, whether active or passive, are not considered in direct relation to something substantial, but to something accidental. This leads to the conclusion that the act of understanding or sensing is not a substantial being, but accidental, which relates to the functions of the intellect and the senses.

Although the generative and nutritive powers are oriented toward the production or conservation of substance, this occurs through the transformation of matter. Thus, their action, like that of other natural agents, takes place through an accidental principle, showing that the powers of the soul operate by mediating accidental principles.

Saint Thomas reaffirms that the essence of the soul is not the immediate principle of its operations. Instead, the soul acts through accidental principles, implying that the powers of the soul are not its essence, but properties of it.

Finally, he concludes that the diversity of the soul's actions, which are of different types and cannot be reduced to a single immediate principle, supports his argument. Since some actions are active and others passive, they must be attributed to different principles. Therefore, although the essence of the soul is a principle, it cannot be the immediate principle of all its actions; it is necessary that the soul possesses multiple powers that correspond to the diversity of its actions.

> Saint Thomas then responds to each of the seventeen arguments initially presented, which claimed that the soul is its powers

1- Saint Thomas clarifies that the book mentioned, *De spiritu et anima,* is not by Saint Augustine, but is attributed to a Cistercian author. He also points out that it is not relevant to worry too much about what is said in that book. If one accepts the idea that the soul is its own power, it could be argued that the powers of the soul are natural properties. Saint Thomas uses the comparison that just as heat, light, and lightness are aspects of fire, the powers of the soul are diverse, but they originate from a single soul.

2- The second, third, and fourth arguments are to be answered in the same way.

5- It is clarified that an accident does not surpass the subject in its existence, but it can do so in its action. For example, the heat of fire can affect external objects. The powers of the soul can exceed it in their ability to understand and love not only themselves but also other things. Saint Thomas refutes Saint Augustine's comparison between knowledge and love in relation to the mind, arguing that this implies the soul could not know or love anything outside itself, which is erroneous.

6- The image of the Trinity in the soul is understood not only in terms of powers but also of essence. This means that, although the powers are distinct, the essence of the soul is unique, just as the divine essence is presented in three persons. If the soul were only its power, there would be no true distinction between the powers.

7- Saint Thomas classifies accidents into three types: those that derive from species, those that depend on the individual, and those that are separable or inseparable. Accidents are not part of the essence of a thing and cannot be defined without considering its essence. This means that one can understand what the soul is without considering its powers, but the soul cannot be conceived without them.

8- It is explained that the distinctions between the sensible and the rational do not directly derive from the sense or the intellect, but from the nature of sensitive and intellectual souls. This emphasizes that the soul possesses capacities that go beyond these distinctions.

9- Saint Thomas refers to the argument previously presented showing why the substantial form does not act as an immediate principle in lower agents.

10- The essence of the soul is the principle of action, but it is a primary principle, not an immediate one. The powers operate thanks to the virtue of the soul, similar to how the qualities of elements act through their substantial forms.

11- The soul itself is potentially capable of intelligible forms. However, this capacity is not the essence of the soul, just as the potential to be a statue does not define the essence of the material from which it is made.

12- Prime matter has the potential to receive the substantial form, meaning its essence is tied to this potentiality.

13- The claim that man is his intellect is interpreted as the intellect being the highest in man. It does not imply that the essence of the soul is only the intellectual power.

14- The similarity between the soul and knowledge is established because both are first acts, but this does not mean the soul is the immediate principle of all operations as knowledge is.

15- The powers of the soul are not essential parts that constitute the essence of the soul, but they are potential parts. The virtue of the soul is distinguished through these powers.

16- Saint Thomas explains that a simple form, which is not subsistent or is pure act, cannot be the subject of an accident. However, the human soul is a subsistent form and is not pure act, which allows it to be the subject of certain powers, such as those of the intellect and will. The sensitive and nutritive powers reside in the body as their subject.

17- Although the soul is unique in essence, it contains both potency and act, allowing it to relate in various ways to things. This ability to relate to the body in different ways allows diverse powers to arise from the single essence of the soul.

13. QUESTION 13: Whether the powers of the soul are distinguished from one another by their objects

> Saint Thomas presents twenty arguments according to which it seems that the powers of the soul are not distinguished by their objects

1- Opposites are what differ the most. However, the fact that two objects (such as white and black) are contrary does not imply that the powers that perceive them (in this case, sight) are differentiated. Therefore, he concludes that there is no distinction in the powers of the soul based on the diversity of objects.

2- A comparison is made between substantial and accidental differences. Man and the stone differ substantially, while sound and color are accidental differences. Since both types of objects relate to the same power, it is argued that the diversity of objects does not cause a difference in the powers.

3- This argument suggests that if the difference of objects were the cause of the diversity in the powers, then the identity of an object should result in the identity of the power. However, the same object can be the object of different powers (for example, what is understood and what is desired). This demonstrates that the difference of objects does not cause a diversity of powers.

4- It is established that if different objects provoke different powers, then the same cause should give rise to similar effects. However, we see that some objects relate to various powers and can also relate to the same power (such as sound and color, which are perceived by both imagination and intellect). Therefore, the difference of objects is not the cause of the diversity in the powers.

5- It is argued that habits are perfections of the powers. Since the powers are distinguished by their habits, they cannot be differentiated according to the objects. This suggests that the classification should be made according to the perfection that each power can reach.

6- It is stated that the powers of the soul exist in the bodily organs, which are the receptors. Therefore, they must be distinguished according to the organs and not according to the objects they perceive.

7- It is held that the powers of the soul are not the essence of the soul itself, but properties derived from it. Since all properties originate from a single essence, there must be a single power that directly derives from the essence of the soul, from which other powers flow in a determined order. This implies that the powers are distinguished by their origin, not by their objects.

8- It is argued that if the powers of the soul are different, one must derive from another. However, all powers exist simultaneously, which implies that there cannot be a hierarchy of origin between them. This strengthens the idea that the powers cannot be distinguished by the diversity of objects.

9- The higher a substance is, the more unified its virtue (internal strength or capacity of a being to perform its acts according to its nature) is. Since the soul is superior to lower beings, its virtue is more unique and is not multiplied by the diversity of objects.

10- If the powers of the soul were differentiated according to the objects, one would expect there to be an order in the powers corresponding to the order of the objects. However, we observe that the intellect (with substantial objects) is posterior to the sense (with accidental objects), indicating that the distinction cannot be simply established by the diversity of objects.

11- Everything that is appetible is either sensible or intelligible. Since the intellect and the sense are the powers that seek their perfection, there is no need to postulate a distinct appetitive power from the sensitive and intellectual ones.

12- The appetite is manifested in the will (intellectual) and in the irascible and concupiscible appetites (sensitive). Therefore, there is no need to consider the appetitive power separately from the sensitive and intellectual powers.

13- It is mentioned that the principles of movement in animals are sense, imagination, intellect, and appetite. Since the powers are those that initiate movement, this indicates that there is no motor power independent of the cognitive and appetitive ones.

14- The powers of the soul are not oriented toward something superior to nature, since the powers attributed to the vegetative soul are directed to natural functions such as the conservation of the species. Therefore, they are not relevant for classifying the powers of the soul.

15- Since the virtue of the soul is higher than that of nature, it is more likely to operate through a single virtue rather than through multiple powers, contradicting the idea that the generative, nutritive, and growth powers are different.

16- It is argued that since the senses are cognitive of accidents and some accidents are more different from one another than others (such as sound and color), if the powers were distinguished by the diversity of objects, they should be even more distinguished by these other accidents.

17- It is posited that each genus has a single principal contrariety. If the sensory powers are diversified according to the various kinds of qualities, one would have to accept that each contrariety involves distinct powers. However, this is not observed in all senses, as in the case of touch.[8]

18- It is argued that memory and sense are not completely separate powers, but that memory is understood as a particular function of the basic sensory capacity. **According to Aristotle, although sense and memory relate to different types of objects**—the sense perceives the present and memory retains the past—**this does not mean that they require two completely independent faculties**.

In simple terms, Aristotle suggests that memory is an extension or a particular use of sense: sense captures an impression of something present, and then memory allows that impression to be preserved over time. Thus, although sense and memory deal with different temporal aspects (present vs. past), both depend on the same basic sensory capacity. Rather than seeing memory as a separate faculty, Aristotle considers it a function of the same sensory power that already enables perception.

19- It is stated that all objects known through the senses are also known through the intellect. If the sensory powers were distinguished according to the plurality of objects, then the intellect would also have to be distinguished into different powers, which is false.

20- Finally, it is mentioned that the possible intellect and the agent intellect are different powers, but both share the same object. This reinforces the idea that the powers are not distinguished by the diversity of objects.

> Then, St. Thomas presents two arguments from authority in favor of the idea that the powers of the soul are distinguished by their objects

1- In this argument, it is stated that the powers (or capacities) of the soul are distinguished through their acts, and that the acts, in turn, are distinguished according to the objects to which they refer. A hierarchy is referenced, in which the powers of the soul (such as sense, imagination, reason, etc.) are defined and differentiated based on the actions or functions they perform, and these actions are determined by the objects these powers know or perceive.

This means that, for example, the power of sight is activated and distinguished from other powers through its act of seeing, which has as its object colors and shapes. Thus, the nature of each power is revealed in how it acts, and these acts are relevant depending on the objects to which they are directed.

2- In this argument, it is asserted that things that can be perfected are distinguished by the perfections they attain. It is said that the objects are the perfections of the powers, implying that the objects toward which these powers tend are, in a certain sense, what defines and determines them.

This means that, if we consider, for example, the object of knowledge (truth) as a perfection of the cognitive power, then the powers would be distinguished according to the different objects they can attain (such as the sense, which is directed to the sensible, and the intellect, which is directed to the intelligible). However, as in the first argument, it is held that although the objects are relevant to understanding the acts of the powers, they are not the cause of the diversity of the powers in themselves.

St. Thomas then offers his own response to the Question raised

St. Thomas responds to the Question of whether the powers of the soul are distinguished from one another according to their objects, affirming that they are, due to the relationship between power and act. He explains that a power is defined in terms of its act, and this act is specified according to its object. Therefore, the diversity of acts, and ultimately of powers, depends on the difference in objects.

He follows a logic in which, if the object of a power is active, it then acts on the power in a passive way (as the objects perceived by the senses). If the object is passive, the power acts on it as an end (for example, in the active powers of the soul). This leads to the conclusion that each act has its own specificity depending on the type of object with which it is related. Hence, the powers of the soul are distinguished according to the different types of objects with which they interact.

He then establishes a classification of three levels of the soul's action: vegetative life, sensitive life, and intellectual life, each having different powers according to their specific functions:

1- Vegetative: Corresponds to the basic life of living beings, encompassing functions such as generation, growth, and nutrition, which are necessary for the existence and conservation of the organism.

2- Sensitive: Involves the powers that allow the perception of objects through the senses and the ability to react to them. This includes the functions of the external senses, such as sight and hearing, as well as internal senses such as memory and imagination.

3- Intellectual: The highest, allowing a person to grasp the essences of things abstractly, beyond material conditions.

This distinction of the powers according to their objects is essential in St. Thomas' theory to understand how the human soul performs different operations based on the nature of the object it faces.

> St. Thomas then responds to each of the twenty arguments that suggest the powers of the soul are not distinguished by their objects

1- Since opposites differ greatly, but belong to the same genus, the diversity of objects according to genus corresponds to the diversity of powers, as genus is, in some way, power. Therefore, opposites relate to the same power.

2- Although sound and color are different accidents, they differ in relation to the change in the sense; on the other hand, man and stone do not, as both affect the sense in the same way. Therefore, man and stone differ accidentally as they are perceived, though they differ in themselves as substances. There is no reason for there to be a difference that is in itself related to one genus and accidentally to another; thus, white and black differ in themselves in the genus of color, but not in the genus of substances.

3- The same thing is compared to different powers of the soul not according to the same reason of the object, but according to other different ones.

4- The higher the power, the more it extends; therefore, it has a more general reason for the object. Thus, some things coincide in the reason of the object of a higher power, which are distinguished in the reason of the object in relation to lower powers.

5- Habits are not perfections of the powers by which they are powers, but as something related to that for which they are, namely, the objects. Therefore, powers are not distinguished according to habits but according to objects; likewise, artificial things are not distinguished according to the objects but according to the ends.

6- The powers are not because of the organs, but rather the other way around; therefore, the organs are more distinguished according to the objects than vice versa.

7- The soul has a main end, such as the intelligible good of the human soul. However, it has other ends ordered to this ultimate end, as the sensible is ordered to the intelligible. And since the soul is ordered to its objects through the powers, it follows that the sentient power is in the human being in relation to the intellective power, and so on with the others. Therefore, according to the reason of the end, one power of the soul arises from another in relation to the objects. Thus, distinguishing the powers of the soul by their origin and by the objects is not contradictory.

8- Although an accident cannot, by itself, be the subject of another accident, the subject, however, subjects itself to an accident through another; thus, the body relates to color through the surface. Therefore, an accident arises from a subject through another, and a power arises from the essence of the soul through another.

9- A soul has, by virtue, a broader capacity than a natural being; thus, sight grasps all visible things. However, the soul, by its nobility, has many more operations than an inanimate being; therefore, it must have more powers.

10- The order of the powers of the soul is according to the order of the objects. But in both cases, the order can be considered either according to perfection, so that the intellect is prior to the sense; or according to the way of generation, and thus the sense is prior to the intellect, because in the way of generation, the accidental disposition precedes the substantial form.

11- The intellect naturally desires the intelligible as such; indeed, the intellect naturally desires to understand, and the sense to feel. But since the sensible or intelligible thing is not only desired to be felt or understood, but also for something else, it is necessary that in addition to the sense and the intellect, there is an appetitive power.

12- The will is in the reason insofar as it follows the understanding of reason; the operation of the will belongs to the same degree of operation as the powers of the soul, but not to the same genus. Likewise, the same is said of the irascible and concupiscible in relation to the sense.

13- The intellect and the appetite move as those who impose the movement; but there must be a motory power that executes the movement, according to which the members follow the command of the appetite, and of the intellect or of the sense.

14- The powers of the vegetative soul are called natural forces because they do not operate except what nature produces in bodies; but they are called the forces of the soul because they do so in a higher way, as mentioned before.

15- The natural inanimate thing simultaneously receives the proper species and quantity; which is not possible in living things, which must have a moderated quantity at the beginning of their generation, because they are generated from a seed. Therefore, in addition to the generative power, there must be an augmentative power that leads to the proper quantity. This must occur through the conversion of something into substance to increase, and thus is added to it. This conversion is carried out by heat, since it also converts what is added externally and resolves what is inside. Therefore, for the conservation of the individual, to continuously restore what has been lost and add what is missing for the perfection of quantity and what is necessary for the generation of the seed, the nutritive power was needed, which serves both the augmentative and the generative; and for this reason, the individual is preserved.

16- Sound and heat and similar things differ according to a different mode of change of the sense, but not so the sensibles of different kinds. Therefore, the sentient powers are not diversified according to those objects.

17- Since the contrarieties of which touch is cognizant are not reduced to a single genus, as the various contrarieties that can be considered about the visible are reduced to a single genus of color, *The Philosopher* determines in the second book of *De Anima* that touch is not a single sense, but several. However, they all agree that they do not feel through an external medium; and they are all called touch, so that it is a single sense divided into several species. However, it could be said that it would be simply a single sense, because all the contrarieties, of which touch is cognizant, are known among themselves and are reduced to a single genus, but it is unnamed; since the proximate genus of hot and cold is unnamed.

18- As the powers of the soul are certain properties, when it is said that memory is the passion of the first sensible, it is not excluded that memory is another distinct power from the sense; rather, its order with respect to the sense is shown.

19- The sense receives the species of the sensibles in the bodily organs and is cognizant of the particulars; however, the intellect receives the species of things without a bodily organ and is cognizant of the universals. Therefore, a certain diversity of objects requires diversity of powers in the sensitive part, which does not require diversity of powers in the intellectual part. To receive and retain in material things is not the same; but in immaterial things, it is the same. Similarly, according to the various modes of change, the sense must be diversified, but not the intellect.

20- The same object, namely the intelligible in act, is compared to the agent intellect as made by it; to the possible intellect, on the other hand, as that which moves it. Therefore, it is evident that it is not compared to both according to the same reason.

14. QUESTION 14: Whether the human soul is corruptible

> Saint Thomas presents twenty-one arguments suggesting that the human soul is corruptible

1- The book of *Ecclesiastes* is cited, which states that there is no difference between the death of humans and that of animals: when animals die, their soul perishes. This implies that, when humans die, their souls could also corrupt, suggesting that the soul is not immortal.

2- It is argued that corruptible and incorruptible things differ in nature. Since the human soul and that of animals do not differ in species, it is concluded that if the soul of animals is corruptible, the same must apply to the human soul.

3- Damascene says that the angel is immortal by grace, not by nature. Since the angel is not inferior to the soul, it is argued that the soul cannot be considered naturally immortal.

4- This argument is based on the notion that the first mover, which is infinitely powerful, moves in infinite time. If it is held that the soul has infinite power, that would imply its essence is also infinite, which is contradictory since only the divine essence is infinite. Therefore, the human soul cannot be incorruptible.

5- An objection is raised to the idea that the soul is incorruptible by divine virtue, arguing that what is not essential to something cannot be considered part of its essence. Since being corruptible or incorruptible is essential to the nature of a being, the soul must be incorruptible by its own essence if it is considered immortal.

6- It is stated that everything that exists is either corruptible or incorruptible. If the human soul is not incorruptible by its nature, it must necessarily be corruptible.

THE HUMAN SOUL

7- It is argued that everything incorruptible has the virtue of being eternal. If the human soul is incorruptible, it should exist always. But this implies there cannot be a state of non-being, which contradicts faith.

8- Saint Augustine is cited, who says that just as God is the life of the soul, the soul is the life of the body. Since death is the deprivation of life, it is concluded that the soul is also deprived and eliminated with death.

9- It is argued that the form (the soul) cannot exist without the body. Therefore, if the body dies, the soul must also perish.

10- The objection that the soul can be corruptible only in its form and not in its essence is answered. It is maintained that the soul is not accidentally the form of the body; it must essentially be the form. Thus, if it is corruptible in its form, it will also be corruptible in its essence.

11- This argument indicates that something constituted by unity is corrupted when one of its elements is corrupted. Since the soul and body form a unity, if the body dies, the soul must also be corrupted.

12- It is established that the sensitive soul and the rational soul are the same essence in the human being. Since the sensitive soul is corruptible, it is concluded that the rational soul is also corruptible.

13- It is suggested that the form (the soul) must be adequate to the matter. If the body is corruptible, then the soul must also be corruptible.

14- It is argued that if the soul can separate from the body, it must have some independent action, but there is no operation of the soul without the body, as knowledge cannot occur without mental images, which depend on the body.

15- It is held that if the human soul is incorruptible, it would only be so because of its ability to understand. However, it is suggested that the activity of understanding is not fully realized, so there is no need to consider the human soul immortal.

16- It is argued that not all humans reach understanding, suggesting that comprehension is not the proper operation of the human soul. Therefore, there is no need to consider the human soul incorruptible.

17- It is maintained that everything finite can be consumed. The natural good of the human soul is finite, and if its goodness is diminished due to sin, it seems that it could eventually be annihilated, implying that the soul can be corrupted.

18- It is argued that the weakness of the body affects the soul. If the body is corruptible, it is suggested that the corruption of the body also implies the corruption of the soul.

19- It is held that everything created from nothing is susceptible to returning to nothing. Since the human soul is created from nothing, it must also be corruptible.

20- It is said that if the cause remains, the effect should also remain. If the soul is the cause of the life of the body, it should remain always. This is false, as we know the body dies.

21- It is concluded that what subsists by itself must belong to a species or genus. Since the human soul is not considered an individual or species, it seems it cannot subsist by itself and therefore cannot exist separately from the body.

> Then, Saint Thomas presents four arguments from authority suggesting that the human soul is not corruptible

1- The human soul is immortal because it is made in the image of God. In the Book of *Wisdom*, it is said that God created man as immortal and in His image. According to Saint Augustine, this image refers to the soul. This argument suggests that since the human soul is created in the image and likeness of God, it shares His divine nature, which implies its incorruptibility. Thus, the soul cannot be destroyed or corrupted because its essence is related to the eternal and divine.

2- The absence of contradictions in the human soul. This argument maintains that everything corruptible must be composed of opposing elements or contradictions. However, the human soul is entirely devoid of contradictions, as even though there may seem to be opposing elements within it, they do not manifest as such within the soul. Therefore, lacking contradictory elements that could lead to its corruption, it is concluded that the human soul is incorruptible.[9]

3- The immaterial nature of the human soul. A parallel is drawn between the human soul and celestial bodies, which are considered incorruptible due to their lack of materiality in the sense of being generable and corruptible. It is argued that the human soul is completely immaterial, as it can receive the species of things without materiality. This immateriality implies that the soul is not subject to the same processes of generation and corruption that affect material bodies, and thus is considered incorruptible.

4- The separation of the intellect from the corruptible body. This argument is based on *The Philosopher*'s assertion that the intellect is perpetual and separate from the corruptible. Since the intellect is a part of the soul, it is deduced that the human soul must also be incorruptible. The idea is that if the intellect, as an aspect of the soul, can exist independently and endure beyond the corruption of the body, then the soul, which includes the intellect, must also be incorruptible.

> Next, Saint Thomas offers his own response to the proposed Question

Saint Thomas argues the incorruptibility of the human soul through several key points that reinforce his thesis. He begins by noting that the being of something is intrinsically related to its form. Every entity has existence (being) by virtue of its specific form; therefore, being cannot be separated from the form that determines it. Composites of matter and form are corruptible because they can lose their form, and by doing so, they lose their being. However, the form itself cannot be corruptible, as it only corrupts accidentally when it loses its relationship with matter. If there were a form that possessed being by itself, it would necessarily have to be incorruptible.

Saint Thomas argues that the human intellect is the faculty that allows a person to understand and know. This capacity for comprehension does not depend on a bodily organ, as there is no organ that can encompass all sensitive natures. Thus, the intellect operates independently of the body, indicating that its existence is not subordinate to material conditions. This fact reinforces the idea that the intellect has a being that transcends the physical realm, implying that it is incorruptible.

The human intellect is not a composite of matter and form but is immaterial, allowing it to receive the species (forms) of things without the limitation of materiality. This is particularly relevant because the intellect can conceive of the universal, which is beyond material conditions. By being immaterial, the intellect also implies that the intellectual principle of the human being is incorruptible.

Saint Thomas concludes that the intellectual principle of man is a form that possesses existence by itself. Since the existence and operation of the intellect do not depend on the body, it is established that the human soul, which includes this intellectual capacity, is incorruptible. He further mentions two signs that support this conclusion: the nature of the intellect, which perceives the corruptible in an incorruptible manner, and the natural appetite of humans for perpetuity, suggesting an intrinsic desire for immortality. Through this logical and philosophical argumentation, Saint Thomas maintains that the human soul is necessarily incorruptible, relying on the relationship between form and being, the nature of the intellect, and the innate human desire for eternity.

> Next, Saint Thomas responds to each of the twenty-one arguments initially presented, which considered the soul subject to corruption

1- On the interpretation of Solomon. It is explained that Solomon in the book of *Proverbs* speaks in different persons, sometimes as a wise person and other times as a fool. The death of humans and animals refers to the corruption of the compound (body and soul), where the separation of the soul from the body causes corruption, although the human soul persists, while the soul of animals does not.

2- Classification of souls. It is argued that if the human soul and the soul of animals were classified in the same way, it could be concluded that they belong to different species. However, since both are parts of a corruptible compound being, they can be considered of the same species.

3- Immutability and mortality. Saint Thomas maintains that true immortality manifests as immutability. Both the soul and angels possess this immutability by grace, which reinforces their incorruptible nature.

4- Being and form. It is clarified that the being is related to the form as a consequence of it. The existence of a being for an infinite time does not prove the infinity of its form, but that of its cause.

5- Essence of corruptibility. Although incorruptibility may belong to the essence, the "perpetual act of existing" (that is, immortality or existence without corruption) does not depend exclusively on the essence of the soul but rather on an external active principle (for example, in theology, this could be interpreted as divine power).

6- General answer. There is no further comment. Aquinas argues that what has already been stated is sufficient to respond to the objection raised.

7- Virtue of the soul. It is noted that the soul has the virtue of existing always, but it has not necessarily always existed. Therefore, it may not have existed in the past, but it will not cease to exist in the future.

8- The nature of the soul. The soul is considered the form of the body inasmuch as it is the principle of life. Life is identified with the existence of the living being.

9- Existence of the soul. The soul has a being that does not depend on the body. This is demonstrated through its operation.

10- Subsistent form. Although the soul is essentially form, it can have characteristics that do not strictly belong to its nature as form, such as subsistence.

11- Unity of the human being. Although the soul and body constitute a human being, the existence of the being comes from the soul. Therefore, even if the body departs, the soul remains.

12- Corruptibility of the sensitive soul. It is clarified that the sensitive soul in animals is corruptible, while in humans, since they share the same essence as the rational soul, it is incorruptible.

13- Relationship between body and soul. The human body is properly constituted for the operations of the soul. Corruption and physical defects are consequences of matter, not of the essence of the soul.

14- Intelligence and phantoms. The assertion that one cannot understand without imagination applies to the present state of life, while there is a different mode of comprehension in the separated soul.

15- Capacity for understanding. Although the human soul does not understand in the same way as higher beings, it does manage to understand in a way that demonstrates its incorruptibility.

16- Common knowledge. Although few achieve perfect understanding, all attain sufficient understanding, as the principles of demonstration are common to the conception of the soul.

17- Sin and nature. Sin eliminates grace but does not alter the essence of a being. Some of the inclination toward grace is lost, but the capacity of nature remains.

18- Weakness of the body. The weakness of the body does not weaken the soul, as action depends on the organ, not on the essence of the soul.

19- Corruptibility and essence. It is argued that what arises from nothing can be reverted to nothing, but this does not imply it is corruptible; rather, it has within it a principle of corruption.

20- Incorruptibility of the soul. Although the soul, which is the cause of life, is incorruptible, the body, which receives life from the soul, is subject to transmutation and thus can experience corruption.

21- Nature of the soul. Finally, it is established that although the soul can exist by itself, it does not have a species by itself, as it is part of a greater species.

15. QUESTION 15: Whether the soul, when separated from the body, is capable of understanding

> Saint Thomas presents twenty-one arguments suggesting that the soul, once separated from the body, is incapable of understanding

1- The operation of the composite (body and soul) does not persist in the separated soul. Understanding is an operation belonging to this union, so understanding cannot exist in the soul apart from the body.

2- Aristotle holds that understanding cannot occur without mental images *(phantasmata)*. These images depend on the senses, which are tied to the body. Therefore, the separated soul cannot understand.

3- Even if it's argued that Aristotle refers to the soul joined to the body, the separated soul cannot understand unless it uses its intellectual capacity. Since Aristotle says understanding depends on images, the separated soul cannot understand, as it lacks access to them without the body.

4- Aristotle compares the intellect to the sense of sight. Just as we cannot see colors without them, the intellect cannot understand without mental images, implying it cannot understand without the body.

5- Aristotle mentions that understanding can be affected by internal factors, such as the heart or natural warmth. These elements are part of the body. Thus, the separated soul cannot understand as it is separated from these factors.

6- If it is claimed that the separated soul understands differently than when united with the body, it contradicts the nature of form and matter. The form (soul) unites with the body to complete its action, which is understanding. If it could understand without the body, this union would be unnecessary.

7- If the separated soul understands, it would do so in a nobler manner than when united with the body. However, this would be detrimental since the soul's good lies in understanding, which would make its union with the body unnatural.

8- The soul's powers are distinguished by their objects. If the separated soul can understand without images, it would need to have different powers, which is impossible since the powers are inherent to the soul's nature.

9- If the separated soul understands, it must do so through some power. The only intellectual powers are the agent intellect and possible intellect, both of which depend on images. Thus, it seems the separated soul cannot understand.

10- Every being has its own operation. If the soul's operation is to understand through images, it cannot understand otherwise, meaning it cannot understand apart from the body.

11- If the separated soul understands, it must do so through some likeness with the known object. However, it cannot understand through its own essence, as this is exclusive to God, nor can it understand through the essence of the known object.

12- Innate species would be useless if the soul could not understand through them while in the body. Species only hold value if they serve for comprehension.

13- Even if it's argued that the soul can understand through innate species, the soul joined to the body is more perfect and therefore must be able to understand better than in its separated state.

14- What is natural to something cannot be completely impeded by its nature. If intellectual species are naturally innate to the soul, union with the body should not prevent understanding through them, which contradicts experience.

15- If the separated soul could only understand through previously acquired species, it would mean that some separated souls, which did not acquire species, could not understand, which is unsustainable.

16- If the separated soul only understands through acquired species, it would only understand what it knew while united with the body. However, it can understand things it had no prior knowledge of, such as punishment or reward.

17- Understanding requires the presence of species in the intellect. If the intellect has species, it can understand. Thus, species do not remain in the intellect after it ceases to understand, meaning it cannot understand after separation.

18- Acquired habits produce acts similar to those from which they were acquired. Species are acquired by contemplating images; therefore, the separated soul cannot understand without returning to the images.

19- It cannot be said that the separated soul understands through the species of a higher substance. The human intellect is designed to receive information from the senses and cannot receive from a higher level.

20- It is not enough for a superior cause to act on something that naturally originates from lower causes. The human soul needs to receive its species through the senses, and it cannot do so solely through the influence of higher substances.

21- Action must be proportional to the subject that receives it. The understanding of higher substances is not suited to the human intellect, as these substances have a more universal and abstract understanding. Therefore, the separated soul cannot understand through species originating from these higher substances.

> Next, Saint Thomas presents three arguments from authority according to which the soul, when separated from the body, is indeed capable of understanding

First Argument

Premise 1. The act of understanding *(intelligere)* is the highest and proper operation of the soul. This means that the capacity for understanding is fundamental to the soul's essence.

Premise 2. If it is concluded that understanding is impossible for the soul when it is separated from the body, it suggests that no other operation could belong to it either. This would imply that, in separation, the soul would lack any function or activity, which would be problematic.

Premise 3. If the soul cannot perform any operation without the body, it is concluded that it cannot exist as a separate being. That is, the existence of a soul that performs no actions is contradictory.

Conclusion. Given that the existence of a separated soul is accepted, it must be accepted that it has the capacity for understanding. This reinforces the idea that understanding is essential, and therefore, the separated soul must possess this capacity.

Second Argument

Premise 1. It is mentioned that those resurrected in the Scriptures retain the same knowledge they had before death. This serves as an example suggesting that memory or understanding is not lost with death.

Premise 2. From this, it follows that the knowledge acquired by an individual during earthly life is not lost after death. This suggests that the soul, even when separated from the body, retains access to what it understood and knew.

Conclusion. Therefore, it is held that the soul can understand through the "species" or representations it acquired while united with the body. This implies that knowledge and the capacity for understanding persist beyond the separation from the body, indicating that understanding is a function of the soul not limited by its separated state.

Third Argument

Premise 1. It is established that there is a similarity between lower realities (corporeal realities) and higher ones (spiritual or intellectual realities). For instance, mathematicians can predict future events by observing similarities in celestial bodies.

Premise 2. It is stated that the soul is superior to all corporeal things. As such, it has the capacity to understand the similarities that exist between physical realities and their intellectual representations.

Conclusion. Since all corporeal realities have a representation within the soul, which acts as an intellectual substance, it is suggested that the soul has the ability to understand all corporeal things, even when separated from the body. This reinforces the idea that the soul's understanding does not depend on its union with the body, but rather is inherent to its nature.

Following this, Saint Thomas offers his own response to the Question posed

In his response, Saint Thomas addresses the question of how the human soul is capable of understanding.

He begins by acknowledging that, in its present state, the soul appears to require the senses to understand the sensible world. This has led to various opinions regarding the nature of this need. On the one hand, some philosophers, such as the Platonists, argue that the senses are not necessary for understanding in itself, but merely help recall what the soul already knows innately. According to this view, the soul possesses prior knowledge that can be awakened by sensory experience, and before uniting with the body, it could access this knowledge unimpeded. However, this stance faces the difficulty of explaining why the soul joins with the body, given that its operation could be limited by this union.

Another perspective, that of Avicenna, suggests that the senses are not necessary for acquiring knowledge but rather for preparing the soul to receive knowledge from an active intellect, a separate substance that provides it with intelligible forms. This view, although more elaborate, also encounters the problem of immediate acquisition of knowledge, as it would imply that any soul, at any time, could access all types of knowledge, which is clearly false.

Saint Thomas thus argues that sensory powers are necessary for the soul to understand, not just accidentally but essentially. The images *(phantasmata)* perceived by the senses act as representations of the objects the intellect needs for comprehension. In this way, the senses allow the intellect to achieve a fuller understanding of things.

However, when considering the possibility of the soul being separated from the body, the difficulty arises of how it can understand without the sensory images it normally utilizes. To resolve this, Saint Thomas proposes that while the human soul has a limited participation in intellectual knowledge, it can receive influence from higher substances (angels) even without sensory images once it is separated from the body. However, he clarifies that knowledge acquired through the senses remains superior in precision and detail.

Finally, he emphasizes that separated souls still retain the knowledge they acquired in life, allowing them to understand effectively, though not as completely as if they had continuous access to sensory images. In this way, Saint Thomas' response balances the need for senses in the acquisition of knowledge with the soul's capacity to understand even after separation from the body, pointing to a richer and more nuanced understanding of the relationship between soul, body, and knowledge.

> Following this, Saint Thomas responds to each of the twenty-one arguments suggesting that the soul separated from the body cannot understand

1- Saint Thomas clarifies that Aristotle does not speak from his own opinion, but in relation to the view of those who hold that understanding involves movement. This implies that comprehension is not necessarily tied to physical movement.

2- Here, it is indicated that Aristotle refers to the intellectual operation of the soul in its state united with the body. In this union, the soul needs images *(phantasmata)* to understand, meaning that the intellect's operation depends on sensory experience.

3- In the current state of the soul's union with the body, the soul does not partake in the higher intelligible species; it only has access to the "intellectual light" that enables it to understand through images. However, once separated, the soul will be able to access those intelligible species without needing external objects.

4- The previous idea is reiterated, emphasizing that after separation, the soul will have more direct access to intelligible realities.

5- Saint Thomas notes that Aristotle speaks from the perspective of those who believe that understanding requires a bodily organ, which would be incompatible with the separated soul's capacity to understand.

6- Here it is clarified that the soul unites with the body through its operation, which is understanding. This does not mean it cannot understand without the body, but rather that, in the natural order, understanding is less perfect in union with the body.

7- This is directly related to the previous idea, indicating that reasoning holds in that the union of the soul to the body is necessary for intellectual activity in this life.

8- It is explained that images *(phantasmata)* only become objects of the intellect when they are rendered intelligible by the "light of the active intellect." Thus, the nature of the formal object does not change, although the material object differs.

9- The operation of the active and possible intellect is distinguished in the state of union with the body, indicating that upon separation, the soul will be able to receive species from higher realities directly.

10- The soul's operation is to understand intelligible realities in act, and this operation is not altered by the fact that intelligible species come from images or other sources.

11- Here it is clarified that the separated soul does not understand things through its essence, but through species received from higher substances, which differs from the Platonist position that believed in immediate essential knowledge.

12- The previous idea is reinforced, stating that knowledge through species from higher substances is exclusive to the soul's separated state.

13- It is argued that if the soul had innate species, it could understand without acquired ones. However, its bodily activity limits its ability to access higher realities.

14- It is repeated that it is not natural for the soul to understand through species received in union with the body; rather, this is possible only after separation.

15- Saint Thomas explains that separated souls can understand through previously acquired species, but also through those they receive after separation.

16- This response relates to the intellect's power to understand, which is limited by its union with the body.

17- It is clarified that intelligible species can exist in the possible intellect in a potential state, needing a stimulus to pass into action, which can occur in various degrees.

18- The intellectual operation is not distinguished by the source of species, as what matters is the object itself and not its material origin.

19- The possible intellect is not designed to receive anything from images, but this does not prevent it from receiving influences from higher realities.

20- Science (knowledge) within the soul is tied to images while united with the body, but upon separation, it may acquire knowledge from higher sources.

21- Although the science (the knowledge) of separated substances is not fully adequate for the human soul, this does not mean it cannot receive some influence from them, albeit not in a complete or perfect manner.

16. QUESTION 16: Whether the soul, when united to the body, can understand separated substances

> Saint Thomas presents ten arguments according to which it seems that the soul, when united to the body, can understand separated substances

1- No form is hindered in reaching its end due to the matter to which it is naturally united. The end of the *anima intellectiva* seems to be to understand the separated substances, which are the most intelligible. Just as the end of every thing is to reach its perfection in its operation, the human soul, therefore, should not be hindered from understanding separated substances due to its union with the body.

2- The ultimate end of man is happiness, which, according to Aristotle in the *Nicomachean Ethics*, consists in the operation of the highest faculty, the intellect, in relation to the noblest object, which would be a separated substance. If man could not reach this end, then his existence would be purposeless, which would be absurd. Therefore, man, even when united to the body, should be able to know the separated substances.

3- Every process of generation reaches a term, for nothing moves infinitely. The operation of the intellect also involves a process, where it passes from potency to act, that is, to knowledge in act. This process cannot continue indefinitely and must reach a term in which the intellect is fully in act, which could not occur without knowing all intelligible things, including the separated substances.

4- It is more difficult for the intellect to abstract concepts from material things, which are not naturally separated, than to understand those that are naturally separated. Since the human intellect, united to the body, can abstract concepts from material things, it should have an even greater capacity to understand separated substances.

5- Just as the sensory perception of intense external objects is limited by the capacity of the sensory organ to bear such intensity, the intellect is not corrupted by intelligible objects, but on the contrary, is perfected by them. Therefore, the more intelligible the object is, the more the intellect can understand it, and the separated substances are the most intelligible of all.

6- The intellect, even when united to the body, abstracts the quiddity or essential nature of things. Eventually, the process of abstraction must reach a quidditative nature that is not a concrete thing with essence, but a pure essence. Separated substances, having no materiality, are essentially pure quiddities, so the intellect should be able to know them.

7- It is natural for the intellect to know causes from their effects. Since separated substances produce effects in sensible and material things (because, according to St. Augustine, angels administer the corporeal by God's command), the intellect should be able to understand separated substances from their effects in material things.

8- The soul united to the body can understand itself. Augustine explains in *De Trinitate* that the mind knows and loves itself, and since the human soul shares the nature of separated substances in terms of intellect, it should be able to understand other separated substances.

9- Nothing exists in vain in reality. Therefore, if separated substances are intelligible in themselves, the human intellect should be able to understand them, since otherwise their intelligibility would be pointless.

10- The intellect, in relation to the intelligible, resembles sight in relation to the visible. Just as sight can perceive visible objects, even though it is itself corruptible, the human intellect should be able to know separated substances, which are incorruptible and fully intelligible in themselves.

> Next, Saint Thomas presents an argument from authority according to which the soul united to the body cannot understand separated substances

Aristotle explains in Book III of *De Anima* that the soul cannot understand anything without resorting to the images or phantoms *(phantasmata)* provided by the senses. These phantoms, or sensible representations, are essential for the process of knowledge in the human soul, for without them intellectual understanding is impossible. However, since separated substances are immaterial by nature, they cannot be represented by phantoms, as they lack a material form that the senses can capture and then transfer to the intellect. Therefore, it must be concluded that the soul, in its state united to the body, cannot understand the separated substances, as it depends on phantoms for understanding, and these entities lack a sensible representation accessible to the human senses.

> Next, Saint Thomas offers his own response to the Question

Saint Thomas responds to this Question by acknowledging that Aristotle promised to resolve it in his treatise *De Anima*, although this solution has not reached us. Therefore, various interpretations have arisen regarding whether the human soul can know separated substances.

Some claim that the soul can know them once united to the agent intellect, which they consider a separated substance capable of naturally knowing such substances. According to this theory, the agent intellect would unite with us as a form that enables this knowledge, just as light makes color visible in the pupil. Others believe that the human soul can know separated substances in a manner similar to how it understands material objects, through principles of philosophy.

Saint Thomas rejects both opinions. He argues that the agent intellect, if it is a separated substance, cannot unite with us in a way that becomes part of our being; otherwise, it would not be a separated substance. Moreover, he criticizes the idea that perfect knowledge of intelligible beings leads to knowledge of separated substances, for the human intellect, being dependent on the senses and mental images (phantoms), cannot understand the nature of separated substances.

In conclusion, Saint Thomas maintains that while the soul is united to the body, it can only know separated substances in an indirect manner: through the images and effects of material beings. Therefore, the knowledge of such substances will be partial and negative, allowing only an understanding of "what they are not" rather than what they truly are.

> Next, St. Thomas responds to each of the ten arguments initially presented, which suggested that the soul, when united to the body, can understand separated substances

1- St. Thomas responds that the natural capacity of the human soul extends to the point of being able to know separated substances. The union with the body does not prevent this possibility of knowledge. Moreover, the ultimate happiness of man, which can be attained by natural means, consists in such knowledge of separated substances.

2- The solution to the second argument follows from what was explained in the first, so no further explanation is necessary.

3- The faculty of the possible intellect progresses continuously, moving from potency to act as its understanding increases. However, the ultimate goal of this progression is to know the divine essence, the supreme intelligible. This knowledge cannot be attained solely by natural means; it requires grace.

4- It is more difficult to "make" separated substances than to simply understand them, if they are the same substances. If they are different, it is not necessary to make them in order to understand them. Furthermore, there may be greater difficulty in understanding certain separated substances than in abstracting and understanding others.

5- Unlike the senses, which can be damaged by intense sensory objects, the intellect is not corrupted by being the receptacle of excellent intelligibles, as it lacks a physical organ susceptible to damage. However, there are intelligibles that exceed the capacity of the human intellect, such as separated substances, whose natural understanding is limited due to the intellect's dependence on abstract species of phantasmata. If the intellect could understand separated substances, it would increase its understanding of other objects, rather than diminishing it.

6- The abstract essences of material things are not sufficient to understand what separated substances are, as they do not provide an adequate comprehension of their nature.

7- Similarly, deficient effects are not enough to fully understand the cause from which they originate, as has been said previously.

8- The human possible intellect does not directly understand itself through its own essence, but through the species it receives from the phantasmata. For this reason, the philosopher says that the possible intellect is intelligible in the same way that other objects are. Nothing is intelligible in potency, but in act, as explained in the *Metaphysics*. Since the possible intellect is in potency in terms of its intelligible being, it can only understand itself through the form that actualizes it, which is the abstract species of the phantasmata. This applies to all the faculties of the soul: acts are known through the objects, faculties through the acts, and the soul through its faculties. Thus, the intellectual soul is known through its act of understanding, but the species derived from the phantasmata is not a form of separated substance that allows it to be known, as happens with the possible intellect.

9- This argument is ineffective for two reasons: first, because intelligibles do not exist "for" the intellects that understand them; rather, they are ends and perfections of those intellects. Therefore, it is not true that an intelligible substance, which is not understood by another intellect, is "superfluous" or lacks purpose. Second, even though separated substances are not understood by our intellect when it is united to the body, they are understood by other separated substances.

10- The species that are perceived by sight can be likenesses of any body, whether corruptible or incorruptible. However, the species that the possible intellect receives from the phantasmata are not likenesses of separated substances; therefore, the same comparison cannot be made.

17. QUESTION 17: Whether the soul, when separated from the body, can understand the separated substances

> St. Thomas presents eleven arguments according to which it seems that the soul, when separated from the body, cannot understand the separated substances

1- Argument from the perfection of the operation: It is argued that a substance is more perfect when it is united than when it is separated, implying that the soul, when united to the body, would be more perfect than when separated. Therefore, if the soul united to the body cannot understand the separated substances, it seems that it could not do so once separated.

2- Argument from nature or grace: It is questioned whether the knowledge of the separated substances by the soul can be achieved by nature or only by grace. If it is by nature, the fact that the soul is united to the body should not prevent such knowledge, as it is natural for the soul to be united to the body. If it is by grace, since not all separated souls have grace, not all could know the separated substances.

3- Argument from the purpose of union to the body: It is proposed that the purpose of the union of the soul to the body is to acquire knowledge and virtues. Since the greatest perfection of the soul lies in the knowledge of the separated substances, if the soul could reach this knowledge only by separating, the union to the body would seem useless.

4- Argument from essence or species: If the separated soul knows a separated substance, it should do so through the essence of that substance or through a species of it. However, the essence of a separated substance is not identical to the separated soul, and one cannot abstract a species from a separated substance, as these are simple. Therefore, the separated soul could not know the separated substances.

5- Argument from means of knowledge: It is argued that knowledge can only come through the senses or the intellect. Since the separated substances are not sensible, they cannot be known by the senses; nor could they be known by the intellect, as it does not concern the singular, and separated substances are singular.

6- Argument from the distance between faculties: The distance between the possible intellect of the human soul and an angel is greater than the distance between the imagination and the possible intellect in man. If imagination cannot understand the possible intellect, then the human possible intellect cannot understand a separated substance.

7- Argument from disposition toward the good and truth: Since some separated souls, such as those of the damned, cannot orient themselves toward the good, it is deduced that their intellects also cannot orient themselves toward the truth. Since the knowledge of a separated substance is a supreme form of truth, this would imply that not all separated souls could know the separated substances.

8- Argument from the proximity to happiness: Philosophers maintain that ultimate happiness consists in knowing the separated substances. If the souls of the damned can understand these substances, it would seem that they are closer to happiness than the living, which is contradictory.

9- Argument from the nature of knowledge among intelligences: According to the *Book of Causes*, one intelligence knows another according to the modality of its own substance. However, it is stated that the possible intellect cannot know itself directly but through species derived from phantasms. Therefore, the separated soul could not know other separated substances.

10- Argument from modes of knowledge: There are two modes of knowledge: one in which knowledge of the prior is reached through the posterior, and another in which the posterior is known from the prior. In the case of separated souls, they could not follow the first mode, as it is based on sensory knowledge. Therefore, the separated soul should know by the second mode, which would mean that the most known realities, such as the divine essence, would be the first to be known. This would contradict the doctrine that the vision of the divine essence is reached only by grace, not by natural means.

11- Argument from the impression of one substance on another: An inferior separated substance can only know another if it receives an impression from the latter. However, the impression of a separated substance on the separated soul is weak and very limited. Therefore, the separated soul could not fully understand the separated substances.

> St. Thomas presents an argument from authority according to which the soul, when not united to the body, can understand or know the separated substances

The argument from authority offered against the eleven previous arguments is based on the principle that "like is known by like" (in Latin, *simile a simili cognoscitur*). This means that for something to be known, there must be an affinity or similarity between the knower and the known.

The conclusion of the argument is that the separated soul, being a *substantia separata* (i.e., a substance that exists independently of the body), should have the capacity to know other separated substances (such as angels or immaterial realities). Since the separated soul shares with these other substances the quality of being independent of the material, the affinity between them should allow the soul to understand them.

> St. Thomas then offers his own response to the Question posed

St. Thomas responds that, according to the teachings of the faith, it is reasonable to affirm that separated souls can know the separated substances, namely, the angels and demons, in whose company they are destined, either for their good or for their ill. It does not seem likely that the souls of the damned are ignorant of the demons, with whom they share company and who are terrifying to them; much less likely is it that the souls of the blessed are ignorant of the angels, whose presence brings them joy. This knowledge of the separated substances by the separated souls is reasonable.

During its union with the body, the human soul is oriented towards lower realities due to its relationship with the body, so its knowledge is completed only through the species obtained from the phantasms. Thus, the soul can only come to know itself and others insofar as it is guided by these species. However, when the soul separates from the body, its orientation is no longer dependent on the inferior, and it is capable of directly receiving the influence of higher substances without the mediation of the phantasms, which will no longer be present. In this way, the soul is actualized through this influence, and thus it comes to know itself directly, by contemplating its own essence, and not indirectly, as it does while united to the body.

The essence of the soul belongs to the genus of intellectual separated substances, and although it occupies the lowest level in this genus, it shares with them the mode of subsistence, since all are subsistent forms. Just as one separated substance can know another by contemplating its own essence, in which a likeness of the other substance is found through the influence received from it or from a common superior cause, so too the separated soul, by contemplating its own essence, can know the separated substances according to the influence received from them or from a superior cause, that is, God. However, this knowledge will not be as perfect as the one that the separated substances have among themselves, since the soul occupies the lowest level among them and, therefore, receives the emanation of intelligible light in a limited way.

> Next, St. Thomas responds to each of the eleven arguments initially presented, which held that the soul separated from the body cannot understand or know separated substances

THE HUMAN SOUL

1- The soul united to the body is, in a certain sense, more perfect than the separated soul, in relation to the nature of the species, but the separated soul has a perfection in the act of understanding that it cannot possess while united to the body.

2- The cognition of the separated soul is discussed in the context of what pertains to it by nature; the cognition of separated substances is natural for the soul in its separated state, but not while it is united to the body.

3- The highest knowledge the human soul can attain during its existence is the understanding of separated substances, but the body enables it to progress towards this knowledge through study and merit.

4- The separated soul does not know the essence of the separated substance, but its species and likeness; the species it receives are influences from the higher realities.

5- Knowledge of the singular does not oppose the intellect, except insofar as it is determined by matter; separated substances can be understood in their essential nature.

6- Imagination and human understanding are more compatible with each other than human understanding and angelic understanding, although both coincide in the realm of the intelligible.

7- The condemned souls are disordered with respect to the ultimate end; they can understand many truths, but not the supreme truth, which is God.

8- The true happiness of the human being lies in the knowledge of God, not in that of creatures; the damned, although they know things that we do not, are farther removed from true happiness.

9- The mode of knowledge of the separated soul of itself is different from the one it has while united to the body, allowing for a clearer understanding of its essence.

10- Separated souls can know more clearly what is familiar to them, but this does not imply that they can see God by his nature or essence.

11- Although the impressions of separated substances on the separated soul are received imperfectly, this does not imply that they cannot know them, but that they do so in an imperfect way.

18. QUESTION 18: Whether the soul, separated from the body, knows all natural things

Saint Thomas presents sixteen arguments according to which it seems that the soul, separated from the body, does not know all natural things:

1- According to Saint Augustine, demons know many things through the experience of a long time, which the soul does not have immediately after separation. Since demons have a more penetrating intellect than the soul, it seems that the separated soul cannot know all natural things.

2- Souls united to bodies do not know all natural things. If, upon separation from the body, they could know all things, it would mean they acquire additional knowledge after separation. Some souls have acquired knowledge in this life, which would imply that after separation they would have duplicate knowledge, which seems impossible.

3- No finite power can comprehend the infinite. Since the essence of the separated soul is finite, it cannot know all the infinite aspects of natural things, such as the species of numbers, figures, and proportions, which are infinite.

4- All knowledge occurs by the assimilation between the knower and the known. But since the separated soul is immaterial, it seems impossible for it to assimilate to natural things, which are material. Therefore, it is unlikely that the soul can know natural things.

5- The possible intellect is similar to prime matter in the sensible realm, since it can only receive one form at a time. Therefore, the separated possible intellect could only receive one type of knowledge and could not know all natural things at once.

6- Things that have different species cannot be similar in species to the same subject. Since cognition occurs by the assimilation of the species, a single separated soul cannot know all natural things, which are of diverse species.

7- If separated souls know all natural things, they would have to possess within themselves the forms that are likenesses of natural realities. If they only knew genera and species, they would not know individuals, which are the fullest manifestation of nature. If they knew individuals, it would imply they would have infinite forms, which is impossible.

8- It is maintained that separated souls have likenesses of genera and species and can apply them to individuals. However, universal understanding cannot apply to the particular that is no longer known. Thus, individuals would remain unknown to the separated soul.

9- Where there is knowledge, there must be an order between the knower and the known. The souls of the damned lack order, as it is said in hell that there is no order, only eternal horror. Therefore, at least the damned souls would not know natural things.

10- Saint Augustine asserts that the souls of the dead cannot know what happens on Earth. Natural things are those that occur here, so the souls of the dead do not know natural things.

11- Everything that is in potency is reduced to act by that which is in act. While the human soul is united to the body, it is in potency regarding many things it can know, but it does not know them all. If after separation it knows all natural things, it must be through something that enables it to know, such as the agent intellect, which cannot make it understand everything it has not previously known.

12- One might argue that the soul does not reduce to the comprehension of all natural things through the agent intellect, but through some superior substance. However, this would not be natural knowledge, but something artificial. The natural action of a connatural agent is necessary, and the agent intellect is the only one that can act naturally in relation to the human possible intellect.

13- If the separated soul is reduced to the comprehension of all natural things, it must be by God or by an angel. However, an angel cannot be the natural cause of the soul itself. Furthermore, it does not seem fitting that the souls of the damned receive such perfection that they know all natural things after death.

14- The maximum perfection of any being in potency is to be reduced to actuality in everything it can be. If the separated soul knows all natural things, it seems that each separated substance should reach its maximum perfection merely by separating from the body, which seems incoherent.

15- Knowledge brings pleasure. If the separated souls know all natural things, one would expect the souls of the damned to experience great joy, which does not seem appropriate.

16- Isaiah is quoted, saying that the dead do not know what the living do. Since what occurs among the living are natural things, this implies that separated souls do not know all natural things.

> Saint Thomas then presents ten arguments from authority according to which the separated soul understands or knows all natural things. To each of these ten arguments, the Angelic Doctor adds, either correcting or expanding the concepts, with sharp observations. To make reading easier, these observations of Aquinas are placed below each argument. In the Treatise, they appear at the end of Question 18

1- Argument from the connection between the separated soul and separated substances. It is argued that the separated soul knows separated substances, and since these substances contain the species of all natural things, it is concluded that the separated soul knows all natural things.

Observes the Angelic Doctor: The separated soul does not perfectly comprehend the separated substance. Therefore, it is not required that it know everything that exists in it through similarity. This implies that the soul's understanding is limited compared to the fullness of the separated substance.

2- Objection to the limitation of understanding. Although it could be argued that one who sees a separated substance does not necessarily see all the species in their understanding, this is countered by Gregory's assertion that those who see God (who sees all things) also see the things that the angels see, implying that those who see angels also understand the things they know.

The Angelic Doctor observes: It is accepted that Gregory's statement about God's knowledge is valid with regard to the intelligible object, since God represents everything that is intelligible. However, it is not necessary that the one who sees God knows everything that He knows unless they fully comprehend it, as God comprehends Himself. This highlights the difference between knowing God and fully understanding what He knows.

3- Intelligibility of the separated substance. It is stated that the separated soul knows the separated substance in its intelligible aspect, which implies that, just as it understands the substance, it must also understand the intelligible species that are in its understanding.

The Angelic Doctor observes: The species that are in the intellect of the angels are intelligible to their own nature, but not necessarily to the intellect of the separated soul. This indicates that the understanding of each being is distinct, and the separated soul cannot access the same level of knowledge as the angels.

4- Identity between the understood and the understanding. According to this argument, what is understood in act is the form of the one who understands. Therefore, if the separated soul understands the separated substance, it can be concluded that it also understands all the natural things that derive from it.

The Angelic Doctor observes: Although it is stated that the understood is the form of the subject who understands, this does not mean that the separated soul, by understanding the separated substance, also comprehends what is in its intellect, as it does not have a complete understanding of that substance.

5- Relation between greater and lesser intelligibles. If someone knows the greater intelligibles, they must also know the lesser ones. Thus, if the separated soul understands the separated substances, which are very intelligible, it should also understand other lesser intelligibles.

The Angelic Doctor observes: Although the separated soul may somehow know the separated substances, this does not imply that it knows all things perfectly, since it does not even fully know the separated substances themselves. This limits its capacity for understanding to an imperfect level.

6- Potentiality and act of understanding. It is established that if something is in potentiality for many things, it is reduced to act through an active principle that has them in act. Since the possible understanding of the soul is in potentiality with respect to all intelligibles, and the separated substance is in act with respect to them, it is concluded that the separated soul understands all natural things.

The Angelic Doctor observes: The separated soul can be led to understand all intelligible things by a superior substance, but this understanding is only universal and not perfect, as mentioned earlier. This indicates that there is a general knowledge, but not a detailed one.

7- Models of the inferior. According to Dionysius, superior beings are models of inferior beings. Since separated substances are superior to natural things, it is inferred that separated souls, by contemplating these substances, know all natural things.

The Angelic Doctor observes: Although the separated substances can be seen as models of all natural things, this does not imply that by knowing these substances, all things are known, unless these substances are understood in their entirety. This reinforces the idea that knowledge is incomplete.

8- Knowledge through infused forms. It is argued that separated souls know things through infused forms, which represent the order of the universe. Therefore, separated souls know the entire order of the universe and, consequently, all natural things.

The Angelic Doctor observes: The separated soul knows through infused forms, but these forms do not specifically represent the forms of the order of the universe, only in a general way. This limits the knowledge of the soul to a more abstract and less concrete level.

9- Presence of the inferior in the superior. Everything that is in the inferior nature is somehow found in the superior. Since the separated soul is superior to natural things and knows itself, it is concluded that it also knows all natural things.

The Angelic Doctor observes: Natural things exist in some way both in the separated substances and in the soul. However, in the separated substances, they are in act, and in the soul, they are in potentiality, meaning the soul has the potential to understand all natural forms, but does not possess them in actuality.

10- Knowledge in the story of Lazarus and the rich man. It is maintained that Luke's narrative about Lazarus and the rich man is not a parable, but a real event. In this story, it is said that the rich man in hell recognized Abraham, whom he did not know previously. From this, it is deduced that separated souls, including those of the damned, can know things they did not know in life, suggesting that they can know all natural things.

The Angelic Doctor observes: It is mentioned that Abraham's soul was a separated substance, and therefore, the rich man's soul could recognize it, as well as other separated substances. This implies that recognition among separated souls is possible, suggesting a capacity for knowledge that may transcend earthly experience.

> Next, St. Thomas offers his own response to the Question at hand

St. Thomas responds to the Question raised by affirming that the separated soul understands all natural beings, but in a relative and not absolute way. To clarify this, he points out that there is an order in things: what exists in the lower nature is also present in a more excellent way in the higher nature. For example, qualities like heat and cold are specific in generated and corruptible beings, but they are manifested universally in celestial bodies. Similarly, the forms of material bodies exist in a particular way in these bodies, whereas in intellectual substances, they exist in an immaterial and universal manner.

St. Thomas mentions that in the universe, everything that exists is found in a more perfect way in God, where forms and natures are united and simple. He explains that in creation, what is manifested is the being of things through the Word of God, and these forms are understood by angelic intelligences, who grasp them in their purity and universality.

The author highlights that the knowledge of the species of things is part of the perfection of understanding, whereas the knowledge of individuals is not to the same extent. Individuals are significant only insofar as they preserve the species that nature seeks to generate. Therefore, the separated soul, although it has a perfect understanding, does not have it to the same degree. Superior substances have more united and universal forms, while inferior ones are more numerous and less universal, approaching particularity.

The intellectual capacity of the human soul, which is the lowest among intellectual substances, is united with the body to receive the forms of material beings. Thus, the human soul can understand intelligible species through its union with the body, while when separated, it can only grasp influences from universal forms. Although this reception is less universal than in superior substances, it lacks the necessary potency to achieve perfect and specific knowledge of each thing, leaving its understanding in a universality and confusion similar to that found in universal principles.

Separated souls acquire this knowledge suddenly, like an influence, and not through gradual learning. Thus, it is concluded that separated souls have universal knowledge of all natural beings, though not in a specific manner for each one. Furthermore, there is a distinction regarding the knowledge possessed by the souls of the saints by grace, which makes them comparable to angels by seeing all things in the Word.

> Next, St. Thomas responds to each of the sixteen arguments by which it seems that the separated soul does not understand or know all natural things

1- According to St. Augustine, demons know things in three ways: by revelation from good angels, by the sharpness of their own intellect, and by prolonged experience. This suggests that their knowledge is limited to certain sources and is not absolute.

2- Those who have acquired knowledge in this life have a determined understanding of what they have learned and a more confused knowledge of other aspects. Therefore, it is not contradictory that both types of knowledge exist in them.

3- The reason presented does not relate to the issue at hand, since it is not claimed that the separated soul knows all natural things in a specific way, which allows for an infinite number of species in numbers and proportions.

4- The forms of material things exist immaterially in immaterial substances, establishing a relationship between them in terms of the reasons for forms, though not in terms of their mode of existence.

5- Prime matter only relates to forms in two ways: in pure potency or in pure act. Natural forms operate immediately when they are in matter, whereas the possible intellect relates to intelligible species in various ways.

6- A cognitive substance can be assimilated in two ways: according to its natural being or according to its intelligible being, allowing various intelligible species to be assimilated to it.

7- Separated souls not only know species but also individuals, though not all of them, meaning that they do not need to contain infinite species.

8- The application of universal knowledge to the singular does not cause the knowledge of the singular, but is rather a consequence of it.

9- Good consists in mode, species, and order; in the damned souls, there is no grace but only natural good, so they do not have the necessary order for this type of knowledge.

10- St. Augustine refers to singulars that occur here and do not belong to intelligible knowledge.

11- The possible intellect cannot reach the knowledge of all natural things solely by the agent intellect, but by a superior substance that possesses complete knowledge.

12- The answer to this argument follows from the previous one.

13- Separated souls receive perfection from God through angels, who transmit not only natural perfections but also those related to the mysteries of grace.

14- A separated soul, with universal knowledge of natural beings, is not perfectly in act, since knowing something in universal terms implies imperfect and potential knowledge.

15- The damned mourn for the knowledge they possess, as they are aware of being deprived of the supreme good to which they were oriented.

16- The gloss refers to particular beings that do not contribute to the perfection of intelligible knowledge.

19. QUESTION 19: Wheter the sentient powers remain in the separated soul

> Saint Thomas presents twenty arguments suggesting that the sentient powers remain in the separated soul

1- The powers of the soul are essential or natural properties. Since nothing can be separated from its essence or natural properties, the sentient powers must remain in the separated soul.

2- If it is said that the sentient powers remain in the separated soul only as in a root (potentially), this implies they are not in act. However, essences and properties must be in act in the substance. Therefore, the sentient powers cannot remain in the separated soul only potentially.

3- Saint Augustine asserts that upon separation from the body, the soul carries with it sense and imagination, which belong to the sensitive part. This suggests that the sentient powers indeed remain in the separated soul.

4- A whole cannot be considered complete if it lacks parts. Since the sentient powers are parts of the soul, if they do not remain in the separated soul, it would not be whole.

5- Humans are defined by reason and intellect, while animals by their capacity to feel. If the sentient powers do not remain in the separated soul, the sense of the resurrected human would not be the same, contradicting the idea of resurrection.

6- Saint Augustine mentions that souls in hell experience visions similar to those of the sleeping, which implies these visions are produced by imagination, a part of the sensitive soul. Thus, the sentient powers exist in the separated soul.

7- Joy and anger are emotions from the concupiscible and irascible appetites. If separated souls experience joy (in the case of the good) and sorrow (in the case of the wicked), the sentient powers must also exist in them.

8- Pseudo-Dionysius the Areopagite describes the evils of the demon as irrational rage and concupiscence. Since these characteristics belong to the sentient powers, it follows that these powers are also present in separated souls.

9- Saint Augustine mentions that the soul can feel without a body, experiencing emotions such as joy and sorrow. This suggests that the capacity to feel exists in the separated soul.

10- The *Book of Causes* states that there are sensitive things in every soul. Since these things are perceived because they are in the soul, it is inferred that the separated soul also has the capacity to feel.

11- Saint Gregory asserts that the story of the rich man Epulon is not a parable, indicating that the separated soul sees and hears. This shows that the sentient powers are in the separated soul.

12- The sensitive and rational soul are the same substance. If one cannot exist without the other, then the sentient powers must remain in the separated rational soul.

13- What is lost in death cannot be identically recovered. If the sentient powers do not remain in the separated soul, they cannot reemerge in the same state, contradicting the resurrection.

14- Divine justice corresponds to human merits and demerits, which are manifested through the actions of the sentient powers. This implies that these powers must exist in separated souls so that they can be rewarded or punished.

15- Power is the principle of action or passion. The soul is the principle of sensitive operations, implying that the sentient powers must remain in the soul since they cannot be destroyed without the destruction of the subject.

16- Memory belongs to the sensitive part. If memory exists in the separated soul, as demonstrated in the story of the rich man Epulon, this indicates that the sentient powers are in the separated soul.

17- Virtues and vices remain in separated souls, and some of them reside in the sensitive part. Therefore, the sentient powers must remain in the separated soul.

18- There have been cases of resurrected persons who claimed to have seen imaginary things (houses, fields, rivers). This suggests that separated souls use imagination, which belongs to the sensitive part.

19- The senses contribute to intellectual knowledge. If the capacity to sense is perfected in the separated soul, this suggests that the sentient powers must also be present in it.

20- Aristotle indicates that an elderly man who receives a young eye would see as a young person. This suggests that the sentient powers are not affected by the weakness of the organs, so they would not disappear at death, implying that they remain in the separated soul.

> Following this, Saint Thomas presents four arguments of authority suggesting that the sentient powers do not remain in the separated soul

1- Argument of perpetuity and corruption. *The Philosopher* states that only that which is perpetual can be separated from what is corruptible. The sentient powers, being functions of the soul dependent on corporality, cannot exist separately, as their nature is intrinsically related to matter. Therefore, since they are not perpetual, the sentient powers cannot remain in the soul once it separates from the body.

2- Dependence on the operations of the sentient powers. *The Philosopher* establishes that the operations of powers requiring a body cannot exist without it. The sentient powers operate through bodily organs, and if these operations cannot be carried out without the body, it follows that the powers themselves cannot exist without it. Consequently, the sentient powers cannot remain in the separated soul.

3- Proper operations of the powers. Saint John Damascene teaches that nothing can be devoid of its proper operation. If the sentient powers remained in the separated soul, they would have to possess their own operations. However, since these powers depend on corporality to operate, their existence in the separated soul would be contradictory and therefore impossible.

4- Frustration of power. It is argued that it would be absurd for a power to exist without being able to act, as there is no room for frustration in the works of God. The sentient powers, requiring a body to act, could not remain in a separated soul where they cannot carry out their functions. The lack of capacity for action would mean these powers would be unnecessary and useless in the context of the separated soul, contradicting the perfection of divine work.

Saint Thomas then offers his own response to the Question at hand

In this text, Saint Thomas argues that these powers are not essential to the soul itself but are natural properties that derive from its essence. He maintains that the sentient powers are not part of the soul's essence but are properties that flow from it. This means that the powers are related to the nature of the soul but are not its fundamental substance.

It is explained that an accident, such as the powers, can be corrupted in two ways: by the action of a contrary or by the corruption of its subject. Since the soul's powers do not have an inherent contrary, they can only be destroyed when their subject, which is the body, is corrupted. This establishes that if the sentient powers are destroyed, it is only through the corruption of the body.

To understand whether the sentient powers continue to exist in the separated soul or are corrupted with the body, it is essential to consider what constitutes the subject of these powers. Saint Thomas points out that the subject must be something that can act or suffer; that is, it must be an entity capable of action or passion.

Additionally, Saint Thomas discusses various views on sensory operations. The Neoplatonists, for example, argued that the sensitive soul has its own operation and can move itself, separating itself from action on the body. They propose a distinction between internal operations, which are proper to the soul, and external ones, which depend on the body. However, Saint Thomas refutes this position, arguing that the sentient powers cannot operate independently of the body and therefore cannot have separate existence.

Finally, he concludes that the sentient powers exist in the composite—that is, in the union of the soul and body—as in their subject, but they derive from the soul's essence as a principle. When the body is destroyed, the sentient powers are also destroyed, although they remain in the soul as their root or principle. Therefore, the sentient powers are not independent of the body and cannot remain in the separated soul in an active way.

> Saint Thomas then responds to each of the twenty arguments suggesting that the sentient powers remain in the separated soul

1- The sentient powers are not part of the soul's essence. Saint Thomas establishes that the sentient powers are natural properties that depend on the soul as a principle but are not essential to its existence. This means that, although the powers derive from the essence of the soul, they are not part of its fundamental constitution.

2- The powers in the separated soul are like roots. It is argued that the sentient powers in the separated soul exist potentially, not actually. The separated soul can be seen as capable of manifesting these powers again if it reunites with a body, similar to how a plant can grow from its root.

3- Authenticity of an authority cited. Saint Thomas rejects the validity of a quote attributed to Saint Augustine, arguing that the text in question is not truly his. He suggests that the quote could be interpreted to acknowledge the powers in the separated soul not as existing, but as potential.

4- Nature of the soul's powers. It is clarified that the powers of the soul are not essential parts but are potential. Some powers are inherent to the soul itself, while others are present in the composite of the soul and body.

5- Differentiation of sense. A distinction is made between sense as a principle of the sensitive soul and sense as a power. In the human context, sense as the essence of the sensitive and rational soul is the same, so the resurrected human maintains their identity, although the properties and accidents may not be the same.

6- Retraction of a position on hell. Saint Thomas refers to Saint Augustine's retraction on the nature of Hell, suggesting that any argument related to Hell should also be reconsidered in light of this retraction.[10]

7- Emotions in the separated soul. It is explained that in the separated soul, emotions such as joy or anger do not exist in their sensitive form, as these are inherent to the sensitive part. However, there can be movements of the will, which belong to the intellectual part.

8- Evil and demons. It is discussed that evil in demons should not be understood in terms of sensitive emotions, but rather as qualities that correspond to their intellectual nature, using the term "evil" in an analogous way.

9- Senses without bodily organs. Saint Thomas clarifies that assertions that the soul can feel without a body do not imply sensory perception without organs but refer to feelings like fear and sadness, which do not depend directly on interaction with physical objects.

10- Incorporation of objects in the separated soul. It is argued that sensitive objects are not in the separated soul in a sensitive way but are perceived in an intelligible manner.

11-Metaphors in scripture. It is noted that some descriptions in the Gospels are metaphorical, such as the visions of Lazarus, implying that it should not be taken literally that the separated soul perceives through the senses.

12-Substance of the soul after death. It is asserted that the essence of the sensitive soul persists after death, but the sentient powers do not.

13-Relationship between powers and body. Thomas Aquinas compares the sentient powers to sense perception, affirming that they are not acts of the body itself but depend on the soul as their principle, and that resurrection does not require a new sensory organ.

14-Reward and merit. It is explained that reward in the afterlife does not require all actions to be reconstituted, but rather that they are remembered by God, avoiding the need to relive painful experiences.

15-The Soul as principle of perception. Emphasis is placed on the soul as the principle of perception, not as an entity that feels, but as the principle that allows sensation through the powers.

16-Memory in the separated soul. Memory in the separated soul is not the same as in the sensitive part but belongs to the intellectual part, considered an aspect of the image of God.

17-Virtues and vices in the separated soul. It is argued that the virtues and vices associated with the irrational part of the soul do not remain in the separated soul, except for their principles in the will and reason.

18-Knowledge of the separated soul. It is established that the knowledge of the separated soul is not the same as that of the soul united to the body, as the latter requires imagination. The separated soul has its own way of knowing.

19-Dependence of the intellect on sense. It is clarified that the intellect needs sense only in a state of imperfect cognition, not in the perfection that corresponds to the separated soul.

20-Weakening of powers. Finally, Thomas Aquinas concludes that the sentient powers are not weakened in themselves, but their corruption is an indirect effect of the corruption of the organism to which they are attached.

20. QUESTION 20: Whether the soul, separated from the body, knows individual beings

> St. Thomas presents eighteen arguments according to which it seems that the soul separated from the body does not know individual beings

1- In the separated soul, only the intellect remains as a power of the soul. However, the object of the intellect is the universal, not the singular. Science refers to the universal, while the senses refer to the singular, as Aristotle says in *De Anima*. Therefore, the separated soul does not know the singular, but only the universal.

2- If the separated soul knows the singular, it will do so through forms acquired previously while it was in the body, or through forms that are infused into it. However, it cannot know through previously acquired forms, since some of these forms are individual intentions that remain in the powers of sense, which cannot subsist in the separated soul. The universal intentions that reside in the intellect are the only ones that can remain, but these do not allow knowledge of the singular. Thus, the separated soul cannot know the singular through the species it acquired in the body. The same is true for infused species, because such species relate equally to all singulars. This would imply that the separated soul would know all singulars, which does not seem to be the case.[11]

3- The knowledge of the separated soul is hindered by the distance of place. St. Augustine, in his work *De cura pro mortuis gerenda*, affirms that the souls of the dead cannot know what happens here. However, the knowledge that comes through infused species is not affected by distance. Therefore, the soul does not know the singular through infused species.

4- The infused species relate equally to the present and the future, since the influence of the intelligible species is not subject to time. If the separated soul knows the singular through infused species, it would seem that it not only knows the present or the past, but also the future, which cannot be, since knowing the future is exclusive to God, as mentioned in *Isaiah* 10, 22.

5- Singulars are infinite, while infused species are not infinite. Therefore, the separated soul cannot know the singular through infused species.

6- The indistinct cannot be the principle of distinct knowledge. However, the knowledge of the singular is distinct. Since infused forms are indistinct (universal), it seems that the separated soul cannot know the singular through them.

7- Everything received in something is received according to the nature of the receiver. The separated soul is immaterial, so the infused forms are received immaterially. However, the immaterial cannot be the principle of knowledge of the singular, which is individuated by matter. Therefore, the separated soul cannot know the singular through infused forms.

8- One might argue that the separated soul can know the singular through infused forms, because they are likenesses of the ideal reasons, by which God knows both the universal and the singular. However, God knows the singular insofar as it is the principle of individuation, which is matter. The infused forms of the separated soul are not productive matter, since that belongs only to God. Therefore, the separated soul cannot know the singular through infused forms.

9- The similarity of the creature to God cannot be by univocation, but only by analogy. But the knowledge given through the analogical similarity is very imperfect; it is as if something were known through another, to the extent that both share existence. If the separated soul knows the singular through infused species, it would seem to do so in the most imperfect way.

10- It has been said before that the separated soul does not know the natural through infused forms, but in a confused and universal way.

However, this cannot be considered knowledge. Therefore, the separated soul does not know the singular through infused species.

11- The infused species through which it is said that the separated soul knows the singular are not immediately caused by God, since, according to Dionysius, the divine law is to reduce the inferior through the intermediate. Nor are they caused by an angel, because an angel cannot cause such species, since it is not the creator of anything. Therefore, it seems that the separated soul does not have infused species through which it knows the singular.

12- If the separated soul knows the singular through infused species, it can only be in two ways: either by applying the species to the singulars, or by turning towards the species themselves. If it applies the species to the singulars, it is clear that such application does not occur by taking something from the singulars, since it does not have sensory powers that could receive from them. Therefore, the application occurs by placing something in relation to the singulars, and thus it will not know the singulars themselves, but only what it places in relation to them. If it knows the singulars by turning towards the species, it will follow that it will only know them as they are in those species. However, in the mentioned species, there are no singulars, but only universals. Therefore, the separated soul does not know the singulars, but only in a universal way.

13- No finite being can know the infinite. Singulars are infinite. Therefore, since the power of the separated soul is finite, it seems that it cannot know the singular.

14- The separated soul cannot know anything except through intellectual vision. However, St. Augustine, in his *Super Genesim ad Litteram*, says that through intellectual vision, neither bodies nor likenesses are known. Therefore, since the singulars are bodies, it seems that they cannot be known by the separated soul.

15- Where there is the same nature, there is the same mode of operation. The separated soul has the same nature as the soul united to the body. Since the soul united to the body cannot know the singular through the intellect, it seems that the separated soul cannot do so either.

16- The powers are distinguished according to their objects. But what each one receives, it receives in a distinct manner. Therefore, the objects are more distinct than the powers. However, sense will never become intellect. Therefore, the singular, which is sensible, will never become intelligible.

17- The cognitive power of a higher order is less multiplied with respect to the knowable of the same order than the power of a lower order. The common sense knows everything perceived by the five external senses. Similarly, the angel, with a single cognitive power, the intellect, knows both the universal and the singular that man perceives through sense and intellect. However, the power of a lower order can never grasp what belongs to another order that is distinct from it, just as the sense of sight cannot grasp what is the object of hearing. Therefore, the intellect of man will never be able to grasp the singular, which is the object of sense, even though the intellect of the angel can know both.

18- In the *Book of Causes*, it is said that the intelligence knows things insofar as it is their cause or governs them. However, the separated soul neither causes the singulars nor governs them. Therefore, it does not know them.

> Next, St. Thomas presents three arguments from authority, according to which the soul separated from the body knows particular beings. To each of these three arguments, the Doctor Angelicus adds, either correcting or expanding upon the concepts, sharp observations. To facilitate reading, we place these observations from Aquinas below each argument. In the text of the Treatise, they appear at the end of Question 20

Argument 1: Formation of propositions

The act of forming propositions is an exclusive function of the intellect. When the soul, even while united to the body, forms a proposition with a singular subject and a universal predicate (for example, "Socrates is a man"), it implies that the soul must know the singular and its relationship with the universal. This demonstrates that the intellect, when operating with singularities, has the ability to know particular beings. Therefore, the separated soul also has the ability to know singularities through the intellect, as the formulation of propositions is evidence of this cognitive ability.

The Doctor Angelicus observes: The soul united to the body can know the singular not directly, but through a reflection. That is, when understanding what is intelligible, the soul reflects on its own activity and the intelligible species that gives rise to its operation. From this reflection, it can consider the images (phantasms) and the particulars, since images are representations of the singular. However, this reflection requires the activity of the imagination and the cogitative power, which are not present in the separated soul. Therefore, the separated soul cannot know the singular in this way.

Argument 2: Comparison with the angels

The human soul is inferior in nature to all the angels. However, the angels of the lower hierarchy receive illuminations about the singular effects, unlike the angels of the higher hierarchies, who focus on the universal reasons for the effects. Since particular cognition is more intense in beings of a lower order, it is inferred that the separated soul, being of a lower order than the angels, has an even greater capacity to know particular beings. This reinforces the idea that the separated soul has access to the cognition of singularities.

The Doctor Angelicus observes: The angels of the lower hierarchy know the reasons for the singular effects not through individual species, but through universal reasons. This is due to their great intellectual capacity, which allows them to understand the singular from the universal. Although the reasons they perceive are universal in themselves, they are considered particular in comparison with the more universal reasons received by the angels of the higher hierarchy. This implies that the angels have a broader and deeper knowledge than the separated soul.

Argument 3: Capacities of the soul in relation to the senses

Anything that a lower power can do, a higher power can also do. Just as the sense, which is a faculty lower than the intellect, can know particular beings, it can be concluded that the separated soul, by functioning through its intellect, can also know singularities. This argument highlights the idea that, since the senses can grasp the singular, the intellect of the separated soul must be capable of performing this task even more effectively.

The Doctor Angelicus observes: It is stated that what a being of lower hierarchy (in this case, the sense) can do, a being of higher hierarchy (the intellect) can also do, but in a more excellent way. That is, the same things that the senses perceive in a material and singular way, the intellect understands in an immaterial and universal way. This difference in the mode of knowledge emphasizes the superiority of intellectual understanding over sensory perception and shows that intellectual comprehension is more complete than sensory perception.

Next, St. Thomas offers his own response to the Question raised

St. Thomas holds that the separated soul can know certain singularities, though not all of them. This cognitive ability is related to the memory of what it knew while united to the body, which allows it to recall past experiences and prevents the soul from being devoid of consciousness. Furthermore, the separated soul can know singularities even after the separation, as otherwise, it could not experience suffering, such as the fire of Hell and other bodily pains. However, it does not have knowledge of all singularities through its natural knowledge, which is evident in the fact that the souls of the deceased are ignorant of what happens on earth, as St. Augustine indicates.

The Question presents two difficulties: the first is common to all, since the human intellect seems limited to the knowledge of universals, which raises doubts about the ability of God, angels, and separated souls to know the singular. Some have even denied that God and the angels have this knowledge, which is unacceptable, as it would imply that divine providence would not apply to things, and that divine judgment over human actions would be annulled. Others suggest that God and the angels know the singulars from the knowledge of the universal causes that govern the order of the universe, as there is nothing in the singulars that does not derive from these universal causes. However, this notion is not sufficient for a true understanding of the singular, since even by uniting universals, one never reaches the singular. For example, when speaking of a "white and musical man," one cannot conclude that it refers to a specific individual, as many people can share those characteristics.

Although some argue that angels and separated souls acquire their knowledge of particulars directly from those very particulars, this idea is incorrect. The reason is that there is a significant difference between the intelligible and the material or sensible; the form of a material thing is not immediately apprehended by the intellect but requires multiple intermediaries. The form of a sensible object must pass through different levels before reaching the intellect, which makes it impossible for angels or the separated soul to directly know the forms of particulars.

Instead, the forms that allow the intellect to know are of two types: some are causes of things, and others are received from them. The former allow knowledge based on their generative capacity, while the latter cannot lead to knowledge of particulars, since the artisan knows the house in general terms but cannot identify it without the help of the senses. For His part, God, through His intellect, not only produces the form that accounts for the universal but also the matter that is the principle of individuation, which allows Him to know both the universal and the particular.

Saint Thomas concludes that separated entities, such as souls, can know not only universals but also particulars, as the intelligible species that emanate from God allow them to perceive things according to their form and matter. This does not imply that the human soul has the same level of knowledge. Its knowledge of particulars is more limited and conditioned by its connection to the body, as its capacity for knowledge is proportional to the universal forms it receives. Thus, the separated soul does not know all natural things in specific and complete terms, but in a confused universality. However, it can know certain particulars to which it has a special relationship or inclination, depending on the impressions it received during life. This shows that the separated soul has the ability to know particularities, though in a partial and not exhaustive manner.

> Saint Thomas answers each of the eighteen arguments that suggest the separated soul does not know particular beings

1- The human intellect knows through species that are abstracted from matter, which prevents it from knowing particulars, which depend on matter. However, the intellect of the separated soul has forms that allow it to know the particular.

2- The separated soul does not know the particular through previously acquired species, but through new species it receives. Nonetheless, this does not imply that it can know all particularities.

3- The separated soul is not limited by physical distance to know, but it does not have the sufficient capacity to know all particulars through the forms it receives.

4- Even angels do not know all future contingents, as they depend on the species of the beings present in their causes. Thus, what does not yet exist in the present cannot be known by them.

5- Angels know natural particulars through a single species, unlike separated souls, which cannot know all particulars.

6- If species were merely received, they could not properly represent the particular. The species the separated soul receives are ideal and can effectively represent the particular.

7- Although the species the separated soul receives are immaterial, they are similar to the things from which they proceed and can distinguish the particular.

8- Intellectual forms do not create things, but they are similar to those that create them in terms of their capacity to represent reality.

9- The forms the separated soul receives are not the same as the ideas in the divine mind, but this does not prevent things from being known through them.

10- The species the soul receives are determined by its disposition, allowing it to know certain particularities.

11- The species in the separated soul are caused by God through the angels, which does not prevent some souls from being superior to certain angels in glory.

12- The separated soul knows the particular to the extent that it is a representation of the particular, and the applications mentioned are not the cause of this type of knowledge.

13- Although the particularities are infinite in potency, they are not in act. Therefore, both angels and separated souls can know infinite particulars one by one, similar to how our mind understands infinite numbers.

14- Saint Augustine does not claim that bodies are not known by the intellect, but rather that the intellect is not moved by bodies in the same way as the senses are.

15- Although the separated soul is of the same nature as the soul united to the body, its separation allows it more access to higher realities and knowledge through intellectual forms.

16- The particular does not become intelligible through sensory modification, but through the representation that an immaterial form can offer.

17- The separated soul receives intellectual species in a way that allows it to know both through sensation and intellect.

18- Although the separated soul does not cause or govern things, it has forms that are similar to those of a cause. This relates to the knowledge it has of reality.

21. QUESTION 21: Whether the soul, separated from the body, can suffer the punishment of corporeal fire

> Saint Thomas presents twenty-two arguments according to which it seems that the separated soul cannot suffer punishment from corporeal fire

1- It is argued that nothing suffers unless it is in potency. Since the separated soul is only in potency according to understanding and does not have sentient powers, it cannot suffer through corporeal fire, as this could be perceived as a pleasurable experience from the standpoint of intellect.

2- It is stated that for something to suffer, there must be communication in the matter between the agent and the patient. Since the soul is immaterial and fire is material, they cannot communicate, so the soul cannot suffer from it.

3- It is said that what does not touch cannot act. Fire cannot touch the soul, even in terms of the ultimate quantity, since the soul is incorporeal. Therefore, it cannot suffer from fire.

4- A distinction is made between suffering as a subject (as wood does with fire) and suffering as an opposite. The soul cannot suffer from fire as a subject, because this would mean the fire would become the form of the soul, which is impossible.

5- It is established that there must be some proportion between the agent and the patient. Since the soul and fire belong to different kinds, there is no proportion, which means the soul cannot suffer from fire.

6- It is claimed that everything that suffers moves. The soul does not move because it is not a body, so it cannot suffer.

7- It is asserted that the soul is more dignified than a body of quintessence, and since the latter is impassible, by greater reason, so is the soul.[12]

8- It is mentioned that the agent is more noble than the patient. Since fire is not more noble than the soul, it cannot act upon it.

9- It is argued that while one could say that fire acts as an instrument of divine justice, it is not suitable for punishing the soul, since it does not align with the nature of fire.

10- It is mentioned that God, being the author of nature, does not act against nature. Making the corporeal act upon the incorporeal would be contrary to nature.

11- It is argued that God cannot make the contradictory true simultaneously. Since impassibility is essential to the soul, it cannot suffer.

12- It is stated that each thing acts according to its own nature. Fire does not have the power to act upon the spiritual, and if God gave it this ability, it would cease to be corporeal fire.

13- It is argued that what is done by divine virtue has a true nature. If the soul suffers from fire, it would have to do so according to the nature of suffering, which would imply that it could receive in an incorporeal way, which would not be a punishment.

14- It is said that no instrument acts instrumentally without exerting its own action. Since fire cannot act upon the soul naturally, it cannot be an instrument of divine justice.

15- The idea that fire stops the soul is refuted. If the soul were united to fire, it would imply that the soul could give it life, which is impossible.

16- It is argued that what is tied to something cannot separate from it. However, condemned spirits sometimes separate from the hellish fire, which implies that they do not suffer in this way.

17- It is maintained that what is tied to something prevents its operation. The soul's proper operation is understanding, and it cannot be impeded by a connection with something corporeal.

18- If the soul's suffering were solely due to restraint, then other corporeal things should be able to cause it more suffering than fire.

19- It is mentioned that according to Augustine and Damascene, the hellish fire is not material, which supports the idea that the soul cannot suffer from corporeal fire.

20- It is said that a servant is punished to be corrected. However, the damned in hell are incorrigible, so they should not be punished with corporeal fire.

21- It is argued that punishments are the consequence of contrary actions. Since the soul has submitted itself to corporeal things, it should not be punished by corporeal things.

22- Finally, it is established that just as rewards are given to the righteous, they are spiritual and not bodily. Therefore, if corporal punishment is mentioned, it should be interpreted metaphorically.

> Next, Saint Thomas presents an argument from authority suggesting that the separated soul can suffer punishment from corporeal fire

This argument is based on the authority of the Scriptures, specifically the passage from the *Gospel of Matthew*, chapter 25, where it is mentioned that both the bodies of the damned and their souls, along with the demons, are punished with the same eternal fire.

1- Identity of the fire. It is noted that the fire used to punish the bodies of the damned is the same as that which is mentioned in relation to the souls and demons. This fire is described as "prepared for the devil and his angels," implying that it has a punitive function for all the beings involved.

2- Necessity of the punishment. According to the argument, it is necessary for the bodies of the damned to be punished with corporeal fire. If it is established that fire has the capacity to inflict pain or suffering on bodies, logic suggests that, since the souls of the damned are also subject to this punishment, they must suffer in a similar manner.

3- Parallel reasoning. An analogy is used, establishing that just as bodies are punished with fire, separated souls should also be, as all belong to the same category of beings under the judgment of divine justice. This equivalence suggests that if bodies can suffer, souls also have the capacity to experience pain through corporeal fire.

> Next, St. Thomas offers his own response to the Question raised

St. Thomas, when addressing the issue of how the soul can suffer punishment due to corporeal fire, states that there have been various opinions on the matter. Some, like Origen, maintain that the fire spoken of in Scripture is only a metaphor to express the spiritual affliction of the soul. However, St. Thomas finds this interpretation insufficient, relying on the idea that, according to St. Augustine, it is necessary to understand that the fire is truly corporeal, as both the bodies of the damned and the souls and demons are punished by it.

Others have argued that although the fire is corporeal, the soul does not suffer directly from it but rather suffers from a kind of imaginary vision of the fire, similar to how a person may become distressed by a terrifying dream that does not reflect physical reality. However, St. Thomas rejects this position because it has already been shown that sensory powers, such as imagination, do not exist in the separated soul. Therefore, he concludes that the separated soul does indeed suffer because of corporeal fire, but the nature of this suffering is complex.

Some argue that the soul suffers by seeing the fire, based on the idea that perceiving it causes suffering. However, St. Thomas observes that seeing is a perfection, and every vision should be pleasant, which contradicts the idea of suffering. Hence, it is suggested that the suffering of the soul comes from perceiving the fire as harmful. This leads to the need to consider whether fire is truly harmful to the soul. St. Thomas concludes that indeed, corporeal fire is harmful to the soul.

The suffering of the soul does not occur through alteration, as it does with bodies, but is manifested through deprivation of what naturally belongs to it. Thus, suffering can occur in two ways: one is through a direct alteration, like a body being burned; the other is through the obstruction of its natural inclination. In the case of the soul, which is not bound to a physical place by its own nature, being bound to a body or corporeal fire is itself a form of suffering, as this goes against its nature and desires.

This type of suffering occurs through the action of a superior force that compels the soul to remain bound to something corporeal. Thus, the soul can be subjected to punishments through corporeal fire not because it is affected physically, but because this forced union implies a limitation of its freedom and spiritual nature. St. Thomas concludes that the greatest suffering of the damned comes from their separation from God and their subjugation to corporeal nature, which is an experience profoundly painful for the soul created to unite with God.

> Next, St. Thomas responds to each of the twenty-two arguments initially presented, according to which it seems that the separated soul cannot suffer punishment from corporeal fire

1- Argument on the reception of fire. In relation to arguments 1 through 7, St. Thomas clarifies that it is not claimed that the soul suffers from corporeal fire merely by receiving it or from the alteration it could cause. This implies that suffering does not occur in the usual physical sense, as the soul has a nature distinct from material bodies.

2- The action of fire as an instrument. In relation to the eighth argument, it is noted that fire acts not by its own virtue, but as an instrument of divine justice. Therefore, what is relevant is not the dignity of the fire itself, but the authority of God's justice that uses it to carry out the punishment.

3- Bodies as instruments of punishment. In relation to the ninth argument, it is established that bodies are suitable instruments for punishing the damned. This is based on the idea that those who refused to submit to God, their superior, must be subjected to inferior creatures as part of their punishment.

4- Actions of God upon nature. In relation to the tenth argument, it is maintained that although God does not act against nature, He can operate on nature, doing things that nature itself cannot. This suggests that the suffering of the soul may be outside the laws of nature.

5- Impassibility of the soul. In relation to the eleventh argument, it addresses that the soul is impassible regarding changes that bodies can cause. This means that, by its essence, the soul does not suffer changes as bodies do.

6- Instrumental action of fire. In relation to the twelfth argument, it is emphasized that fire cannot act upon the soul naturally, but only instrumentally. This means that fire does not lose its nature when acting upon the soul.

7- Modes of suffering. In relation to the thirteenth argument, it is reaffirmed that the soul does not suffer from corporeal fire in the aforementioned ways, reaffirming the focus on using fire as an instrument of divine justice.

8- Relationship of fire to the soul. In relation to the fourteenth argument, it is mentioned that although fire does not heat the soul, it has an operative or aptitudinal relationship to it, which is similar to the connection between bodies and spirits.

9- Union of the soul with fire. In relation to the fifteenth argument, it is clarified that the soul does not unite with fire as a form that gives it life, but in a manner in which spirits connect with corporeal places through the action of virtue.

10- Appreciation of fire as harmful. In relation to the sixteenth argument, it is explained that the soul can be afflicted by fire insofar as it perceives it as harmful, even if it is not physically trapped, which can lead to distress even in the absence of direct contact.

11- Limitation of freedom. In relation to the seventeenth argument, it is stated that although the soul is not hindered in its intellectual operations, it loses some of its natural freedom by being forced to suffer.

12- The punishment of Gehenna. In relation to the eighteenth argument, it is emphasized that the punishment of Gehenna is not limited to the souls but also affects the bodies, with fire being a symbol of the most intense bodily suffering.

13- Interpretation of Augustine. In relation to the nineteenth argument, it is mentioned that St. Augustine does not establish a definitive doctrine on the matter, but rather explores the idea that the suffering of souls may be related to fire as harmful in the context of the detention and binding it exerts on the soul, preventing it from uniting with God.

14- Correction of Gregory's view. In relation to the twentieth argument, the idea that all of God's punishments are purgative is criticized, and it is clarified that there are punishments that lead to final condemnation, showing that punishments can be both corrective and punitive.

15- Contradiction of punishment. In relation to the twenty-first argument, it is asserted that punishment is contrary to the sinner's intention, who seeks to satisfy his own will, while the punishment, coming from divine wisdom, aims to reverse that will.

16- Reward and punishment of the soul. Finally, in relation to the twenty-second argument, a distinction is made between how the soul is rewarded by enjoying what is above it and punished by being subjected to what is below it, suggesting that rewards are spiritual and punishments, bodily.

BY WAY OF AN EPILOGUE

1-What does Saint Thomas ask about the soul in the First Article?
Saint Thomas asks whether the soul can be considered a form and, in turn, if it can exist by itself.

2-How does Saint Thomas define the concept of "individual" in the context of substance?
Saint Thomas points out that an "individual" in the genus of substance is something that can subsist by itself and is complete in some species and genus of substance.

3-What opinion does Saint Thomas criticize regarding the nature of the soul?
He criticizes views that consider the soul as a harmony or a complexion, stating that these concepts do not allow the soul to subsist by itself or to be complete in any species of substance.

4-What role does the vegetative soul play according to Saint Thomas?
The concept of the vegetative soul, according to Saint Thomas Aquinas, refers to one of the three aspects of the soul that he describes in his philosophy. In his view, the soul is not a single, simple entity but is distributed into different "potencies" or faculties that correspond to different types of living beings. These are: 1. The vegetative soul: specific to plants and living beings that carry out basic biological functions such as nutrition, growth, and reproduction. 2. The sensitive soul: specific to animals, which enables sensory perceptions and movement. 3. The rational soul: specific to humans, capable of reasoning and self-awareness.

In this context, when Saint Thomas says that the vegetative soul requires a principle that transcends the active and passive qualities of vegetative functions, he is suggesting that there is something beyond the physical or material properties that make these functions possible (such as the processes of nutrition or growth). Vegetative functions, while biological, cannot be explained solely by material aspects (such as chemical interactions), but require an immaterial or spiritual principle that "organizes" and makes them possible. This principle is understood as a vital principle, something that gives order and purpose to biological functions. In other words, while the physical qualities of vegetative organisms allow for growth and nutrition, it is the vegetative soul that "supervises" and gives coherence to these functions, ensuring they are carried out correctly.

The idea that <u>it transcends the active and passive qualities</u> means that the vegetative soul is not simply a consequence of the material interactions of substances, but rather the ordering and structuring cause of those processes, something that goes beyond mere physical reactions.

5-Why is the idea that the sensitive soul is merely a combination of material qualities unsustainable?
Saint Thomas holds that the sensitive soul performs operations that cannot be explained solely by material qualities, as it receives species without matter.

6-What is the relationship between the intellect and matter in Saint Thomas' conception?
According to Saint Thomas, the intellect operates independently of a bodily organ, indicating that its action is distinct from material functions.

7-How does Saint Thomas argue for the intellect's independent existence?
Saint Thomas argues that the intellect must have an existence independent of the body, as its operation does not depend on matter.

8-What concept does Saint Thomas introduce to explain that the human intellect is not merely a part of the soul?
He introduces the idea that the intellect is a substance in itself and does not perish, emphasizing its immortal nature.

9-What criticism does Saint Thomas make of the idea that the human soul is only a part of the body?
He criticizes this idea by affirming that the soul is what gives life to the body and that its separation implies substantial corruption.

10-How does Saint Thomas conclude regarding the relationship between the soul and the body?
He concludes that the human soul is the form of the body, capable of subsisting by itself, although it is not a complete species in itself but completes the human species.

11-What answer does Saint Thomas offer to the Question posed about the nature of the soul?
Saint Thomas answers that the soul is a form that acts independently, subsists by itself, and is essential for the existence of the body, thus completing human nature.

12-What does Saint Thomas ask in the Second Article about the human soul?
Saint Thomas asks whether the human soul, in terms of its act of existing, is separated from the body.

13-What is Saint Thomas' answer regarding the existence of the possible intellect in relation to the body?
Saint Thomas argues that the possible intellect must be considered in potency to what it can know, and as it can understand forms of all things, it cannot be bound to a sensitive nature, which implies it does not have a bodily organ.

14-What does it mean that the possible intellect is "stripped" of sensitive forms?
It means that the possible intellect must be free from all sensitive forms to be able to receive and understand intelligible forms; just as the pupil is devoid of colors to perceive all colors.

15-Why does Saint Thomas criticize the idea that the possible intellect is a form or virtue mixed with the body?

He criticizes this idea because he argues that if the possible intellect were a form or virtue related to the body, it could not perform operations independent of matter, which contradicts its nature.

16-How are the *phantasmata* related to the possible intellect?

Saint Thomas argues that *phantasmata* (mental images) are necessary for the possible intellect to know, but that the possible intellect itself is independent and should not rely on the *phantasmata* for its existence.

17-What error do some make when considering the nature of the possible intellect as separated from the body?

Some believe that the possible intellect is a separate substance that exists independently of the body and can know all intelligible forms, which Saint Thomas refutes, arguing that such a position is incompatible with the fact that a particular individual can know.

18-How does Saint Thomas demonstrate that the possible intellect cannot be a separate substance?

He demonstrates that if the possible intellect were a separate substance, it would be impossible for a specific human being to know through it, as the action of the intellect could not be the action of a principle that does not belong to that particular being.

19-What conclusion does Saint Thomas reach about the nature of the possible intellect?

Saint Thomas concludes that the possible intellect is not a separate substance but rather a capacity of the human soul that, although it is united to the body, allows the human being to perform intellectual operations.

20-What answer does Saint Thomas offer to the second Question posed about the soul-body separation?

Saint Thomas answers that the human soul is a form that not only joins with the body but also possesses an intellectual capacity that manifests independently of material conditions, thus ensuring its existence as a principle of knowledge.

21-What does the Third Article address?
In the Third Question, Saint Thomas asks whether the possible intellect is unique to all men or if there is a possible intellect in each person. He analyzes whether this intellect is a substance separate from the body or if it must be present in each human being individually.

22-According to Saint Thomas, what does the answer to whether there is a common possible intellect for all men or one for each man depend on?
It depends on whether the possible intellect is a substance separate from the body. If it is, then the possible intellect should be unique, as things separate from the body cannot be multiplied by the diversity of bodies.

23-Why does it seem impossible that there is a single possible intellect for all men?
It seems impossible because the possible intellect is the basis for acquiring knowledge, and sciences or knowledge are not the same for all people: some possess knowledge that others lack. This implies that, if the possible intellect were unique, all men would necessarily have the same knowledge, which is absurd.

24-What is the problem with the idea that the possible intellect is unique and that different knowledge depends on the phantasms (mental images) of each person?
The problem lies in the fact that intelligible species or forms cannot be understood without being abstracted from the phantasms and present in the possible intellect. The diversity of phantasms cannot be the cause of the unity or multiplication of knowledge in the intellect, as knowledge depends on intelligible species, which are abstract and universal, not merely on individual phantasms. In other words, the human intellect has the ability to abstract the essences of things and to know them in a unified manner, independently of the sensible representations that each person may have.

25-What problem arises if we accept that the possible intellect is unique in all men?

A difficulty arises because the act of understanding comes from the possible intellect, and if it is unique in all, it would be impossible to explain how different people can simultaneously understand individually and distinctly. This would result in the act of understanding being unique and the same for all, which is impossible.

26-Why is it more reasonable for each person to have an individual possible intellect?

It is more reasonable for each person to have their own possible intellect because the act of understanding is a proper and specific operation of each individual. If everyone shared a single possible intellect, all humans would have the same intellectual nature and operation, which would eliminate individual diversity in understanding and be incompatible with human nature.

27-How does Saint Thomas explain the individuation of the possible intellect in each person?

He explains that the possible intellect multiplies according to the number of human individuals, due to the union of the soul with a specific body in each case. Although the human soul does not completely depend on the body for existence, its union with a particular body allows the multiplication of individual souls without changing the species.

28-How does the individuation of the human soul differ from the individuation of other forms?

The individuation of the human soul does not depend on the body, as it is a subsistent form in itself. However, upon uniting with individual bodies, the human soul multiplies in number, though not in species. This characteristic distinguishes it from other forms, which rely on the body for individuality and cannot subsist by themselves.

29-What is the main topic of the Fourth Article?

The Fourth Article addresses whether there exists an agent intellect, and Saint Thomas defends its existence to explain how the process of knowledge functions.

30-Why does Saint Thomas consider it necessary to postulate an agent intellect?
Saint Thomas considers the agent intellect necessary because the possible intellect is in potential concerning the concepts or intelligibles it must understand. The possible intellect needs to be activated by something that is already intelligible in order to know.

31-What role does the agent intellect play in the movement of the possible intellect?
The agent intellect moves the possible intellect so that it understands something. The objects understood by the possible intellect do not exist as independent entities but are understood universally, as common ideas applicable to various individuals.

32-How does the agent intellect help in the abstraction from matter?
The agent intellect abstracts ideas from the material conditions that individualize them, allowing the essence of things to be grasped without being limited to individual particularities (intellection of essences).

33-What difference does Saint Thomas establish between his view and that of the Platonists regarding universals?
Saint Thomas distances himself from the Platonic view that universals exist by themselves in reality. For him, if this were true, an agent intellect would not be necessary, as material objects could directly move the possible intellect. But, disagreeing with this theory, he considers it essential to postulate an agent intellect.

34-How does the possible intellect come to know immaterial substances?
The possible intellect cannot know immaterial substances directly but understands them indirectly through the abstraction performed on material

and sensory objects.

35-Why is the existence of the agent intellect fundamental?
The existence of the agent intellect is fundamental because it enables the human intellect to understand abstract concepts and realities, facilitating the abstraction from material and particular conditions that limit knowledge.

36-What question does St. Thomas in the Fifth Article?
It asks whether there exists a separate agent intellect for all men.

37-Why does Saint Thomas argue that the agent intellect is more suitable to be considered a separate entity than the possible intellect?
Saint Thomas argues that the agent intellect is more suitable to be considered a separate entity due to its active and universal nature, which allows it to operate independently of the material and particular limitations of the possible intellect.

38-How is the possible intellect manifested according to Saint Thomas?
The possible intellect is manifested in two states: sometimes in potency and other times in act, depending on whether it is in the process of understanding or has already understood something.

39-What is the fundamental difference between the agent intellect and the possible intellect?
The fundamental difference is that the agent intellect is an active principle that performs the action of understanding, while the possible intellect is an internal capacity of the human being to receive and process information.

40-What does it imply that the agent intellect can operate independently?
It implies that the agent intellect can abstract ideas and concepts without needing to be connected to specific sensory data, thus showing a higher

level of intellectual activity.

41-Why can't the possible intellect be separated from the human being?
The possible intellect cannot be separated from the human being because it is intimately linked to the essence of the human being; its nature is to receive and understand ideas based on sensory experiences.

42-What relationship do some philosophers establish between the agent intellect and separate entities?
Some philosophers consider the agent intellect to be a separate substance, which they call "intelligence," and which relates to human souls similarly to how higher substances relate to the souls of celestial bodies.

43-How does God relate to the agent intellect according to Catholic teaching?
Catholic teaching holds that God is the only one who acts in our souls, and Saint Thomas argues that the agent intellect cannot be considered God, as this contradicts its role as a source of knowledge.

44-What kind of active principles does the human being require for intellectual operations?
The human being requires a particular active principle, which in this case is the agent intellect, as opposed to universal active principles that affect all lower bodies.

45-What are the implications of considering the agent intellect a separate entity from God?
Considering the agent intellect as a separate entity would imply that human perfection and happiness would depend on union with something other than God, which contradicts the evangelical teaching on eternal life as the knowledge of God.

46-Why is it impossible for the agent intellect to be a separate substance?

It is impossible for the agent intellect to be a separate substance because its operations require an intrinsic formal principle that cannot be external, as is the case with the possible intellect.

47-How are mental images related to the possible and agent intellects?
Mental images (phantasms) are in potency regarding the entities they represent, and the possible intellect is in potency for all intelligibles, but it is determined to understand through abstract species.

48-What analogy does Saint Thomas use to describe the activity of the agent intellect?
Saint Thomas compares the activity of the agent intellect to a light that makes colors visible, indicating that it abstracts images from their material conditions.

49-What conclusions does Saint Thomas draw about the nature of the agent and possible intellects?
Saint Thomas concludes that both the possible and agent intellects are essential for human understanding and reside within the soul, avoiding theological and philosophical confusions that contradict the Catholic faith.

50-What does the Sixth Question address?
The Sixth Question examines whether the soul is composed of matter and form, questioning the opinions of previous philosophers such as Avicebron.

51-What does Avicebron hold about the soul?
Avicebron argues that, since the soul has properties similar to those of matter, such as being receptive and potential, it must be composed of matter.

52-How does Saint Thomas respond to Avicebron's assertion?
Saint Thomas rejects the idea that the soul is composed of matter and form, considering it frivolous and impossible.

53-What is the difference in the way the soul and matter receive, according to Saint Thomas?
Saint Thomas explains that matter receives with a change or movement, while the soul receives knowledge without undergoing physical transformation.

54-Why does Saint Thomas argue that the soul cannot be a substance composed of matter and form?
If the soul were composed of matter and form, it would create a separate and independent species from the body, contradicting the Aristotelian doctrine that body and soul together form the human species.

55-What does the incompatibility of the soul's composition with its union to the body imply?
If the soul were a combination of matter and form, it could not be the formal principle that gives existence to the body, contradicting its vital role in the body's life.

56-How does Saint Thomas criticize theories about the union of the soul and body that mention "light" or energy?
Saint Thomas considers these ideas "fantastic" and unnecessarily complicating the relationship between soul and body, as he maintains that the soul unites to the body naturally and directly.

57-What does it mean for the soul to be a "subsistent form"?
The soul is a "subsistent form" because, although it has no matter, it exists independently and can subsist without the body.

58-What types of composition does Saint Thomas find in the human soul?
Saint Thomas identifies two types of composition in the human soul: that of essence (*essentia*: what the soul is) and the act of being or existing (*esse* or *actus essendi*).

59-How are essence and the act of being related in the soul?

The essence of the soul has the capacity to exist, but it becomes a real being only when it receives the act of being or existing *(esse)*.

60-What does the structure of act and potency allow in the soul?

This structure allows him to explain how the human soul can exist without depending on a body, as its essence is completed by uniting with the act of being.

61-What is Saint Thomas's conclusion about the composition of the soul?

Saint Thomas concludes that the soul is a subsistent form that may have composition of act and potency but not of matter and form, as the latter is restricted to material beings.

62-What is the subject of the Seventh Article?

The subject of the Seventh Article is whether the angel and the soul are of different species.

63-What opinion is mentioned about the relationship between the human soul and angels?

It is mentioned that some say that the human soul and angels are of the same species.

64-Who is cited as the first to propose this opinion?

Origen is cited as the first to propose this opinion, as he sought to avoid the errors of the ancient heretics.

65-What is Origen's argument about the diversity of creatures?

Origen argues that the diversity of creatures arises from free will, not from God's initial creation.

66-How does Origen explain the differences in rational creatures?

Origen holds that all rational creatures were created equal, and that some progressed by adhering to God, while others fell by distancing themselves from Him.

67-According to Saint Thomas, what is lacking in Origen's argument?
Saint Thomas points out that Origen's argument disregards the consideration of the good of the whole in creation and focuses only on the good of the parts.

68-How is the perfection of a creature related to its species, according to Saint Thomas?
Saint Thomas argues that in God's creation not all creatures are equal, as a perfect universe requires different degrees of beings.

69-What distinction does Saint Thomas establish between angels and souls?
Saint Thomas establishes that angels and souls are different in species, as they cannot be considered as forms of the same matter.

70-What does the statement that angels and souls are not of the same species imply?
It implies that there are formal differences between them, as the form is what gives being to the thing.

71-What does Saint Thomas consider about the matter of angels and souls?
Saint Thomas considers that, since angels and souls are not composed of matter and form, their difference cannot be material.

72-What conclusion does Saint Thomas reach regarding the species of angels and souls?
Saint Thomas concludes that it is impossible for angels and the soul to be of the same species, given that there are formal and perfectional differences.

73-How are species classified among material substances?

Among material substances, different species are classified according to the degrees of perfection of nature.

74-What relationship does Saint Thomas establish between degrees of perfection and species in immaterial substances?
In immaterial substances, degrees of perfection determine species differences in relation to the first agent, who is perfect.

75-What does the Eighth Article address?
The Eighth Article examines whether the rational soul, that is, the human soul, should unite to a body with the specific characteristics of the human body.

76-What is the fundamental reason for the rational soul to unite with a body?
Saint Thomas explains that, since matter exists for the form, the human body exists for the rational soul. This is necessary because the human soul does not initially possess intelligible knowledge, as occurs in other superior intellectual substances; instead, it is like a *tabula rasa* (blank slate) that needs to receive knowledge from the external world through the senses.

77-Why must the human body be suited to the needs of the rational soul?
Since the rational soul needs to grasp intelligible forms through the senses, it is essential that the human body be optimally suited for sensation, especially the sense of touch, which is fundamental for sensory perception.

78-Why is the sense of touch so important in human nature?
Touch is the foundational sense of all other senses, according to Saint Thomas, as all sensitivity is rooted in it. If the sense of touch is affected (as during sleep), all other senses are also altered.

79-What is the ideal disposition of the human body regarding the sense of touch?
The organ of the sense of touch should have a balance of qualities, such

as warmth and cold, moisture and dryness, which requires a moderate blend of these elements so that it can perceive these qualities without being altered by them. Thus, the human body, being balanced, is the most suitable for the rational soul.

80-How is perfection reflected in the structure of the human body?
The composition of the human body shows a higher level of perfection within lower nature, as it is the most balanced in terms of its mixture of elements. This balance enables the human being to be optimally equipped for sensory and cognitive activity.

81-How is this perfection manifested in the human brain?
The human brain is designed to facilitate internal sensory functions such as imagination, memory, and cognitive faculty. Therefore, the human brain is proportionally larger relative to the body than that of other animals, and its structure allows humans to maintain an upright posture suitable for intellectual operation.

82-Why is the human body corruptible and subject to limitations such as wear and fatigue?
These limitations were not deliberately chosen but are inherent to matter. The human body, being composed of opposing elements, is subject to these defects due to the necessities of matter. Although it was given the best disposition for its sensory functions, the nature of material elements implies a certain vulnerability.

83-Could God have created a human body free of corruption and defects?
While God has the power to create an incorruptible body, Saint Thomas points out that, in the context of nature, what is considered is what aligns with the nature of things themselves, according to Saint Augustine. Originally, God granted humanity the grace of original justice, by which the body was fully subject to the soul as long as the soul remained united with God. Upon losing this original justice due to sin, man became subject to the inherent defects of matter.

84-What does the Ninth Article concern?
The Ninth Article addresses whether the soul is united to bodily matter through an intermediary. Saint Thomas answers that it does not, as the union of the soul with matter does not require any intermediate form or entity. The form of the soul, being substantial, unites directly with matter, constituting the human being in its entirety.

85-What is Saint Thomas's main argument regarding the union of the soul with the body?
Saint Thomas argues that, since the substantial form is what gives being to matter, there cannot be an intermediary substantial form between the soul and matter. The union between the soul and the body is direct, as the soul's substantial form grants the body its being and specific essence. There is no intermediate form, as some philosophers had suggested, that allows matter to progress through various degrees of perfection.

86-How does Saint Thomas define the relationship between forms and matter in natural beings?
Saint Thomas asserts that forms determine the different degrees of perfection in natural beings. Matter, when united with a form, acquires different degrees of existence: from merely being corporeal to being an animated body and finally a rational being. There is no intermediate form, as the substantial form of the soul provides perfection to the human body in each of its stages, from the material to the spiritual.

87-What distinguishes the soul's form from other forms in material beings?
The form of the soul is distinct because it gives the human being its specific and complete existence as a rational being. While other forms determine the material or vital characteristics of a being, such as those defining the body or life in animals, the rational soul is what provides the complete essence of the human being, from the body to spirituality.

88-What does the Tenth Article concern?

The Tenth Article addresses whether the soul is present in the entire body and in each of its parts. Saint Thomas explores how the soul, as the form of the body, relates to each part and to the body as a whole.

89-How does Saint Thomas explain the union of the soul with the body?

Saint Thomas asserts that the soul does not unite with the body through an intermediate part but joins immediately with the entire body. The soul is the form of both the whole and each part of the body.

90-Why is it necessary for the soul to be present in each part of the body?

The soul's presence in each part of the body is necessary because each part receives its existence and species from the soul, which acts as its form. This ensures that the body is a natural whole and not just a composition of parts.

91-What implication does the assertion that the soul gives being to each part of the body have?

This assertion implies that, with the soul present in each part, it is not possible for something to receive its being and species from a separate form, as that would resemble the position of the Platonists, who held that sensory beings participate in separate forms.

92-How does Saint Thomas define the concept of totality in relation to the soul?

Saint Thomas defines totality in three ways: by quantitative division, by comparison to the essential parts of the species, and by comparison to the parts of virtue or power. Totality in the soul refers to its perfection as the form of the body.

93-How does the soul relate to the operations of each part of the body?

The soul exercises its virtue and power in the body, but it is not distributed equally in each part. Each part of the body is related to different operations of the soul, so the soul's potency regarding those operations manifests in the corresponding parts.

94-What limitation does Saint Thomas mention regarding the soul's totality in relation to action?
Saint Thomas mentions that, due to its superior nature, the human soul can perform certain operations, such as understanding and willing, without needing a bodily organ. However, for other operations that require organs, the soul acts in its entirety in the body, though not in each part.

95-What does Article 11 concern?
In Article 11, Saint Thomas inquires whether the human soul is one and the same substance or if multiple souls exist within the human being. He examines various opinions on whether the soul is a unique substance or if several souls coexist within the human body.

96-What stance does Plato hold regarding the soul?
Plato maintains that multiple souls exist in the human body. According to him, the soul joins the body as a mover but not as a form. In his theory, the soul resides in the body in a manner similar to a sailor in a ship, with multiple movers causing the various actions in the human body, without this hindering the unity of the human being.

97-Why does Plato's view on souls within the body present a problem, according to Saint Thomas?
Saint Thomas points out that, according to Plato, if the soul is only a mover and not a form, true unity of the human being is not achieved, nor that of animals. This would imply that the human being would not be one in an absolute sense, as generation and corruption would not be simple but would depend on the relationship between the soul and the body.

98-What problem arises from considering the sensitive soul and the rational soul as distinct according to Plato?
If the sensitive soul and the rational soul were considered distinct forms, multiple predications would arise concerning a single individual. This

would imply that the unity of the human being would be accidental rather than essential, leading to the conclusion that man would not be a single being in an absolute sense.

99-What conclusion does Saint Thomas reach regarding the number of souls in the human being?
Saint Thomas concludes that the human being has only one soul in terms of its substance, which is rational. This soul is responsible for the sensitive and vegetative functions of the human body, as well as the rational capacity. In summary, the human soul is unique and substantial, encompassing both sensitivity, vegetativity, and reason.

101-How does Saint Thomas explain the relationship between the different powers of the human soul?
Saint Thomas explains that the different powers of the human soul are all rooted in a single essence of the soul. When one power intensifies, it may interfere with the operations of other powers or even cause one power to "spill over" into another. This implies that all powers of the soul are unified in a single substantial essence of the human soul.

102-What does Article Twelve address?
Article Twelve addresses whether the soul is its powers, that is, whether the very essence of the soul is the direct and immediate principle of all its operations, or if, instead, the powers are properties distinct from the essence of the soul.

103-What are the main opinions regarding the relationship between the soul and its powers?
There are two main opinions. Some believe that the soul is its powers, meaning that the essence of the soul directly serves as the principle of all its operations. Others consider that the powers of the soul are properties derived from it, but they do not identify with its essence.

104-How does Saint Thomas define "power"?
Saint Thomas defines power as the principle of an operation, whether

action or passion. Power is that by which something acts or is affected, but it is not the subject that acts or suffers in itself; rather, it is that through which it acts.

105-What example does Saint Thomas use to explain the concept of power?

He uses the example of the "constructive power" in a builder, or heat in fire. The builder has the power to build through his skill, and fire heats due to its heat.

106-Why does Saint Thomas reject the idea that the soul is its own powers?

Saint Thomas rejects this idea because each being acts according to what it is in act. Since not all the soul's actions belong to its substantial essence, the principle of these actions cannot be the essence of the soul itself, but rather distinct powers that mediate between the soul's essence and its various operations.

107-What does the diversity of the soul's operations imply in relation to their principles?

The diversity of the soul's operations, such as perception, understanding, and growth, requires distinct principles. The actions and passions of the soul cannot arise from a single immediate principle because they differ in nature and require specific principles suited to each type of operation.

108-How does Saint Thomas explain the relationship between the soul's essence and its powers?

According to Saint Thomas, the essence of the soul is a single principle and cannot serve as the immediate principle of all its actions. The soul's essence operates through accidental principles — that is, powers — that correspond to the diversity of its operations.

109-What role do the active and passive powers play in the soul?

Active and passive powers do not directly correspond to something substantial, but to something accidental. For example, the intellectual and sentient powers are directed towards operations that are accidental, not substantial.

110-What does Article Thirteen address?
Article Thirteen addresses whether the soul's powers are distinguished from one another by their objects.

111-How does Saint Thomas define power in relation to act?
According to Saint Thomas, power is defined in relation to act, as it depends on act for its definition, and it is through the diversity of acts that powers are differentiated. Acts obtain their species from objects, and therefore the distinction of the soul's powers is due to the distinction of their objects.

112-In what way do objects determine the distinction of the soul's powers?
The distinction of the soul's powers is based on the distinction of their objects, as acts derive their species from objects. Objects can be considered as active for passive powers and as ends for active powers. This distinction in objects also determines the distinction of operations.

113-How does Saint Thomas compare the action of inanimate nature with the action of the soul's powers?
Saint Thomas points out that the action of the soul transcends the action of inanimate nature in two aspects: in the manner of acting and in what is achieved. All action of the soul originates from an intrinsic agent, as a living being moves itself to action, while action in inanimate bodies originates from an extrinsic agent.

114-What distinguishes vegetative powers from other powers of the soul?
Vegetative powers, such as the generative, nutritive, and augmentative, are oriented toward the existence and maintenance of the living being as such, which also occurs in inanimate bodies but through an extrinsic agent. Therefore, vegetative powers are considered natural.

115-How is the capacity for sensation and intellect manifested in the soul?

Sensation and intellect allow the soul to contain all things in an immaterial sense, as the soul becomes all things through sensory perception and understanding. Sensory perception receives the forms of things in their material particularity, while the intellect abstracts them entirely from matter.

116-What are the five requirements for perfect sensory cognition?

The five requirements are: (1) the reception of the species of sensible objects (proper sense), (2) the judgment and discernment of sensible objects (common sense), (3) the preservation of perceived sensible species (imagination or fantasy), (4) the knowledge of intentions not apprehended by the senses, such as what is useful or harmful (natural estimative in animals or cogitative in humans), and (5) the retrieval of previous perceptions for current consideration (memory or reminiscence).

117-Why is the sense of sight considered the highest of the senses?

The sense of sight is the highest and most universal of the senses because it perceives sensible things without an attached material alteration, and the objects it perceives are common to both corruptible and incorruptible bodies.

118-What does Article Fourteen address?

Article Fourteen addresses whether the human soul is incorruptible and immortal.

119-Is the human soul incorruptible according to Saint Thomas?

Yes, according to Saint Thomas, the human soul is incorruptible.

120-Why is it said that the human soul is incorruptible?

The human soul is incorruptible because the form that gives being to something cannot be separated from that being without the compound being corrupted. The human soul, which has a principle of intelligence, is a form that possesses being by itself, making it incorruptible.

121-What is meant by "form that has being"?
"Form that has being" refers to a form that not only gives existence to a compound but also possesses existence in itself, implying that it cannot be separated from it without the compound being corrupted.

122-Why does human intelligence not depend on a bodily organ?
Human intelligence does not depend on a bodily organ because the human intellect can comprehend all sensible natures universally, without being limited to material conditions, demonstrating that its operation is independent of the body.

123-Is the human intellect a material or immaterial principle?
The human intellect is an immaterial principle, as it receives species immaterially and is capable of abstract knowledge, without depending on the material conditions of sensible objects.

124-How is the incorruptibility of the human soul related to the intellect?
The incorruptibility of the human soul is related to the intellect, because the intellect operates by itself, independently of the body. This intellectual principle, which is immaterial, makes the human soul incorruptible.

125-What does Saint Thomas say about those who claim the human soul is corruptible?
Saint Thomas notes that those who claim the human soul is corruptible are mistaken, as they deny fundamental premises, such as considering the soul as a composite of matter and form or as dependent on the body for its operation.

126-What is the sign of the incorruptibility of the human soul?
The sign of the incorruptibility of the human soul can be seen in two aspects: first, in the intellect, which perceives things universally and does not undergo corruption; and second, in the natural appetite, which has a

desire for perpetuity that cannot be frustrated, suggesting that the human soul is incorruptible.

127-Why does the natural appetite of humans suggest the incorruptibility of the soul?
The natural appetite of humans, which seeks perpetuity and eternal being, suggests that the human soul is incorruptible, as this desire cannot be frustrated and is oriented toward being itself, without temporal limitations.

128-What does Article Fifteen address?
Article Fifteen addresses whether the human soul can know separately from the body.

129-What is Saint Thomas's main argument regarding the knowledge of the human soul separated from the body?
Saint Thomas argues that, according to the current state of human nature, the soul needs the senses to know, as sensory knowledge is necessary for intellectual activity. However, when the soul separates from the body, it would not need the senses, as it would be fully prepared to know by itself.

130-What is the Platonic view on the relationship between the soul and the senses in the process of knowledge?
The Platonists hold that the senses are necessary for the soul's knowledge, not directly, but as a means by which the soul remembers what it already knows naturally. Through the senses, the soul is reanimated and remembers the knowledge acquired before its union with the body.

131-According to Saint Thomas, what consequences would follow for the union of the soul with the body if the Platonic opinion were followed?
According to the Platonic view, the union of the soul with the body would seem unnecessary, as the soul could operate perfectly without the body. This would be incompatible with human nature, as it would not be

logical for the union of the soul with the body to impede its proper functions, given that the soul is nobler than the body.

132-How does Saint Thomas refute Plato's position on the acquisition of knowledge?
Saint Thomas refutes the Platonic stance by arguing that science does not arise from participation in separate ideas but is obtained through the senses. If one of the senses is lacking, knowledge of what that sense perceives is also lost, demonstrating that the senses are necessary for knowledge.

133-What does Avicenna propose regarding the role of the senses in knowledge?
Avicenna proposes that the senses are not necessary in themselves for knowledge; instead, they act as a medium to prepare the soul to receive the intelligible species from a separate substance known as the "agent intellect."

134-What is the main difference between Avicenna's view and that of Saint Thomas on the use of the senses for knowledge?
The main difference is that Avicenna holds that the senses are not essential for knowledge but merely prepare the soul to receive intelligible species. In contrast, Saint Thomas considers that the senses are necessary not only for preparing knowledge but for correctly representing the objects of knowledge.

135-How does Saint Thomas explain that the soul can know without the senses when it is separated from the body?
Saint Thomas explains that when the soul separates from the body, it is freed from the influence of the senses and can perceive influences from higher substances without the need for the senses. However, this perception will not be as clear or as determined as that which the soul obtains through the senses when united to the body.

**136-What challenges does the concept of a soul separated from the body present according to Saint Thomas?

The challenge lies in the fact that a separated soul would need a different way of perceiving knowledge since, without the senses, there would be no "phantasms" or sensory representations, raising questions about how the soul could understand without those means.

137-What solution does Saint Thomas propose for the difficulty of the soul's perception when separated from the body?

Saint Thomas proposes that, although the separated soul cannot know with the same clarity as when united to the body, it can perceive influences from higher substances (angels) and know without the need for bodily "phantasms."

138-What distinction does Saint Thomas make between the knowledge possessed by the soul while united with the body and when it is separated?

The knowledge of the soul united to the body is more determined and precise, as it depends on the senses. In contrast, when the soul is separated, it can receive knowledge of higher realities, but not with the same clarity and determination as when it is in the body.

139-How can the knowledge of the soul be perfected when it is separated from the body according to Saint Thomas?

The soul, when separated, can perfect its knowledge if it receives divine or supernatural knowledge, allowing it to fully know the truth, including the direct vision of God, which would not be possible when the soul is united with the body.

140-What does Article 16 pose?

It raises the question of whether the human soul, when united to the body, can know or understand separate substances.

141-What does Aristotle propose on this Question according to Saint Thomas?

Saint Thomas points out that Aristotle promised to address this issue in the third book of "De Anima," but he does not explicitly discuss it in the texts that have come down to us. This led to different interpretations among his followers on how to resolve the issue.

142-What do some followers of Aristotle believe about the capacity of the soul united to the body to understand separate substances?

Some followers propose that the human soul, even when united to the body, can understand separate substances, and they argue that this constitutes the highest human happiness. However, there are various opinions on how this understanding occurs.

143-How do some followers explain the soul's ability to understand separate substances?

Some argue that the soul, through the agent intellect, can comprehend separate substances, but not in the same way that it comprehends other intelligible objects, such as those studied in speculative sciences through definitions and demonstrations. They attribute to the agent intellect the ability to understand separate substances.

144-What is the relationship between the agent intellect and the possible intellect according to followers of this theory?

According to this view, the agent intellect is compared to the possible intellect in a way similar to how form relates to matter. The possible intellect receives intelligibles and, as it receives them, becomes united with the agent intellect, allowing it to understand not only the material but also separate substances.

145-What objections are raised regarding the view of the agent intellect as a separate substance?

Saint Thomas notes that some philosophers hold that the possible intellect is corruptible and therefore cannot understand the agent intellect or separate substances. Others argue that the possible intellect is incorruptible and thus could comprehend both the agent intellect and separate substances. Saint Thomas refutes both positions, arguing that they are impossible or vain, as they contradict Aristotle's intentions.

146-Why does Saint Thomas reject the idea that the agent intellect is a separate substance?

Saint Thomas rejects the idea because, according to Aristotle, the agent intellect must unite with the possible intellect to operate in it formally, as a form. This would make it impossible for two separate substances, such as the agent and possible intellects, to formally operate together. Moreover, the idea that the agent intellect operates through a separate substance does not align with the way the human intellect relates to knowledge.

147-How does Saint Thomas refute the position that the agent intellect could formally unite us to separate substances?

Saint Thomas refutes this position by explaining that although the agent intellect can influence the possible intellect, it cannot be understood formally through a separate substance. The comparison to the sun illuminating is incorrect since the possible intellect would not unite with the agent intellect as the eye unites with sunlight.

148-What is the correct stance on how the human soul can understand separate substances?

Saint Thomas holds that, because the human soul is united to the body and inclined toward phantasms or sensory images, it cannot know separate substances directly. However, it can know them indirectly by recognizing their existence and immortality through the effects they produce in the material world, as when we know a cause through the effects it produces.

149-What is the relationship between human happiness and the ability to understand separate substances according to Aristotle and Saint Thomas?

According to Aristotle, human happiness consists in operating in accordance with perfect virtue, and among intellectual virtues, wisdom is the highest. This wisdom is achieved through the knowledge of separate substances, but knowing all intelligible objects perfectly is not required to achieve happiness. The ability to understand separate substances is part of human happiness, but not in the sense of complete and immediate understanding of them.

150-Why is the position that the human soul can know all separate substances unsustainable?
Saint Thomas considers this position unsustainable because knowing all separate substances fully and directly is impossible for any human in this life, except for Christ, who was both God and man. Furthermore, Aristotle does not require such knowledge to achieve human happiness, reinforcing the idea that understanding all separate substances is not necessary to attain ultimate happiness.

151-What final conclusion does Saint Thomas propose on the question of the human soul's capacity to know separate substances?
Saint Thomas concludes that the human soul, while united to the body, can only come to know separate substances insofar as it can perceive their existence and general characteristics through the effects they produce in the material world. Perfect understanding of separate substances is not possible in the current state of human life, nor is it required for human happiness according to Aristotle.

152-What is Article 17 about?
Article 17 concerns whether the soul, when separated from the body, can understand separate substances.

153-What are separate substances according to Saint Thomas?
In this Question, separate substances refer to Angels and Demons, in whose company the souls of separated men reside, whether good or bad.

154-Is it likely that the souls of the damned ignore the Demons?
It does not seem likely that the souls of the damned ignore the Demons, as these souls are destined for the company of Demons, who are said to be terrible for them.

155-Is it likely that the souls of the blessed ignore the Angels?
It seems even less likely that the souls of the blessed ignore the Angels,

as they rejoice in the company of the Angels.

156-How is it that separated souls can know separate substances?
It is reasonable for separated souls to know separate substances, as, being separated from the body, their vision is no longer directed toward lower things, as is the case for souls united to the body, which only know what they receive from phantasms. Once separated, the soul can receive influences from higher substances without depending on phantasms.

157-How will the separated soul know itself?
The separated soul will be able to know itself directly by contemplating its own essence, without the need to rely on phantasms, as it does when united to the body.

158-To what type of separate substances does the essence of the human soul belong?
The essence of the human soul belongs to the category of separate and intellectual substances, although it is the lowest in this category, as all separate substances are subsistent forms.

159-How can separated souls know other separate substances?
Just as one separate substance can know another through influence received from it or from a higher cause, the separated soul can also know other separate substances through influence received from them or from a higher cause, namely, God.

160-How does the knowledge of the separated soul compare with the knowledge that other separated substances have of each other?
The separated soul will not know separate substances as perfectly as other separate substances know each other, as the soul is the lowest of these substances and receives the emanation of intelligible light less perfectly.

161-What is Article 18 about?
Article 18 is about whether the soul, separated from the body, knows all

natural things.

162-How does Saint Thomas understand the knowledge that the separated soul has of natural things?
Saint Thomas explains that the separated soul understands natural things in a relative manner, that is, in a universal mode, but not in a particular or detailed way.

163-How is the relationship between natural things ordered?
Saint Thomas asserts that everything found in lower nature is present in a more excellent way in higher nature. Thus, particular qualities found in lower nature, such as heat and cold, appear in a more universal way in celestial bodies.

164-What is the difference between the knowledge that corporeal substances and intellectual substances have?
Corporeal substances have particular and material forms, while intellectual substances have immaterial and universal forms, enabling them to know the essence of things more generally and less particularly.

165-How are forms in God, according to Saint Thomas?
Saint Thomas teaches that in God, the forms of things exist in a simple and unitary manner, unlike creatures, where forms and natures are multiplied and divided.

166-What is the relationship between the knowledge of things by intellectual substances and the knowledge of things by God
Saint Thomas says that the knowledge of things in intellectual substances is more perfect than that of lower creatures, but still does not reach the perfection of God's knowledge, who has perfect comprehension of all things.

167-How is the knowledge of species in nature obtained?
Saint Thomas holds that true intelligible knowledge is related to species (understood as the forms or universal essences existing in objects), as the

human intellect, or intellectual substances in general, can know the universal essences of things. However, this knowledge pertains more to general and universal forms than to concrete and particular individuals.

168-What does the perfection of intelligible knowledge imply?
The perfection of intelligible knowledge consists in the ability to know the universal essences of things, that is, the general principles underlying particular individuals. Superior intellectual substances know these forms in a more universal, unified, and direct way, while inferior ones perceive them in a more dispersed and particular manner.

169-How is human soul knowledge characterized when united to the body?
When the soul is united to the body, its knowledge is limited to receiving the intelligible species of material objects, according to its intellect's capacity, and it depends on bodily senses to attain knowledge.

170-What happens when the human soul is separated from the body?
When the soul is separated from the body, it no longer receives species from material objects but has a direct knowledge of higher realities, although this knowledge is still less universal and perfect than that of higher intellectual substances.

171-How does the knowledge of separated souls compare with the natural knowledge of lower creatures?
Separated souls know natural things in a universal, but not particular, way, as their intellectual capacity is not as powerful as that of higher intellectual substances. Their knowledge is more general and confused, unlike the precise knowledge possessed by higher creatures.

172-How do separated souls acquire knowledge?
Separated souls acquire knowledge by immediate influence, not gradually or by instruction, as proposed by Origen. This knowledge is acquired suddenly, by receiving the influence of higher realities.

173-How does the knowledge of separated souls differ from the knowledge that saints have through grace?
The knowledge of separated souls is natural and limited to the universal, whereas the knowledge of saints is of a supernatural order, as by grace, they are allowed to see all things in the Word of God, giving them a more complete and direct vision.

174-What is Article 19 about?
Article 19 is about whether the sentient powers remain in the separated soul, that is, if after death, when the soul separates from the body, the sensitive faculties continue to exist.

175-What are the powers of the soul, according to Saint Thomas?
Saint Thomas explains that the powers of the soul are not part of its essence but are natural properties that flow from it.

176-How are accidents or properties corrupted?
Accidents are corrupted in two ways: by their contrary, as cold is destroyed by heat, or by the corruption of their subject. Accidents that have no contrary are destroyed only by the destruction of the subject in which they reside.

177-What happens to the soul's powers when the body is corrupted?
Saint Thomas points out that, since the soul's powers have no contrary, if they are corrupted, it can only occur through the corruption of their subject, namely, by the destruction of the body. Therefore, the soul's sentient powers do not remain once the body has been destroyed.

178-What is the subject of the sentient powers, according to Saint Thomas?
The subject of the powers is that which has the capacity to act or receive action. In this case, the body is the subject of the sentient powers, as sensitive actions depend on the interaction between the soul and the body.

**179-What did philosophers think about the operations of the

sensitive part of the soul?
Plato believed that the sensitive soul had its own operations and that the soul could move itself, moving the body only insofar as it was moved. According to Plato's followers, there were internal operations that occurred within the soul and external operations that happened when the body was moved.

180-Why does Saint Thomas refute the position of Plato's followers?
Saint Thomas refutes this position because he holds that if the sensitive soul had its own operations, it would also have its own subsistence and would not be corrupted with the death of the body. This would imply that the souls of animals would be immortal, which is impossible. Therefore, sensitive operations cannot be independent of the body.

181-How is the relationship between the soul and the sentient powers explained in the composite being?
Saint Thomas explains that the sentient powers belong to the composite being, that is, to the animated body, but they depend on the soul as the principle. The soul does not operate directly; rather, through it, the body performs the sensitive functions. Therefore, the composite being sees, hears, and feels, but through the soul.

182-What happens to the sentient powers of the soul once the body is destroyed?
Once the body is destroyed, the sentient powers of the soul also cease in their action but remain in the soul as their root, as a potential principle, although they no longer operate in the state of separation.

183-What is the subject of Article 20?
Article 20 addresses whether the separated soul can know singular beings.

184-What does Saint Thomas affirm about the knowledge of singular beings by the separated soul?
Saint Thomas affirms that the separated soul knows some singular

beings, but not all.

185-What type of singular beings does the separated soul know?
The separated soul knows those singulars that it knew previously while it was united to the body, and some others that it knows after the separation.

186-Why is it necessary for the separated soul to remember the things it did in life?
It is necessary so that the "worm of conscience" does not disappear in the separated soul, which implies a memory of past actions.

187-How does Saint Thomas explain that the separated soul can suffer bodily punishments in hell?
The separated soul knows some singulars after the separation from the body, which allows it to experience bodily punishments in hell.

188-Does the separated soul know all singular beings in its natural knowledge?
No, Saint Thomas explains that the separated soul does not know all singular beings in its natural knowledge.

189-What does Saint Thomas mean by the "common difficulty" regarding the knowledge of the singulars?
The common difficulty is that the intellect seems to be cognitive only of universals and not of singulars.

190-How does Saint Thomas justify that God and the angels know the singulars?
He justifies that God and the angels know the singulars through their knowledge of universal causes and the universal order.

191-Why did some thinkers argue that God and the angels do not know the singulars?
They argued this because the intellect, in its natural function, seems oriented only to the knowledge of universals.

192-What is the problem with knowing the singulars through the knowledge of universal causes?
The problem is that, even though universal causes may be known, this is not enough for the true knowledge of the singulars, as they do not simply arise from the combination of universals.

193-What example does Saint Thomas use to explain the insufficiency of the knowledge of universals to know singulars?
He uses the example of knowing the order of the stars to predict eclipses, stating that this does not allow one to know a particular eclipse.

194-What solution does Saint Thomas propose for the separated soul and the angels to know the singulars?
He proposes that the intelligible forms derived from God are likenesses of things both in form and in matter, which allows the knowledge of the singulars.

195-Why is it not possible for the separated soul to know the singulars directly from the material things?
Because there is a great distance between the material and the intelligible, and the forms of things cannot pass directly into the intellect of an immaterial being.

196-How does Saint Thomas conceive that the separated substances know the singulars?
He considers that they know the singulars through intelligible forms emanating from divine wisdom, which are representations of things in form and matter.

197-How do angels differ from the separated soul in their knowledge of the singulars?
Angels possess a capacity for knowledge proportional to the universal forms in them, allowing them to know all singulars within the species. In contrast, the separated soul has a limited capacity for knowledge and does

not know all singulars completely.

198-What specific elements can the separated soul know according to Saint Thomas?
It can know those singulars to which it has a particular inclination, such as those that affect it or leave impressions and vestiges in it.

199-Why is the knowledge of the separated soul limited to certain singulars?
Because the knowledge is determined by the receptive nature of the soul, which has a mode of receiving forms based on its inclination or experience.

200-What conclusion does Saint Thomas offer regarding the capacity of the separated soul to know the singulars?
He concludes that the separated soul can know some singulars, but not all, and its knowledge depends on its particular relation or impression with those singulars.

201-What is the subject of Article 21?
Article 21 asks whether the separated soul can suffer punishment from corporeal fire.

202-What is the opinion of some regarding the soul's punishment by fire?
Some maintain that the soul does not suffer punishment from corporeal fire, but that its spiritual affliction is metaphorically represented as fire in the Scriptures. This was the opinion of Origen.

203-Why is this explanation insufficient according to Saint Thomas?
It is insufficient because, according to Saint Augustine, it must be understood that the fire is corporeal, as it is also said that this fire affects the bodies of the damned, as well as demons and souls.

204-What is the second opinion regarding the soul's punishment by corporeal fire?

The second opinion asserts that the fire is corporeal, but the soul does not suffer directly from it; rather, it suffers from its similarity in an imaginary vision, as in dreams where one suffers from seeing something terrifying, even though it is not real.

205-Why does Saint Thomas reject this second opinion?
Saint Thomas rejects it because the sensitive faculties, including imagination, do not remain in the separated soul.

206-Then, how does the separated soul suffer the punishment of fire?

Saint Thomas concludes that the separated soul suffers from the same corporeal fire, although it is difficult to pinpoint exactly how this suffering occurs.

207-What does Saint Gregory say about how the soul experiences fire?
Saint Gregory mentions that the soul suffers from the fire by seeing it, although Saint Thomas questions this explanation, as seeing is generally something pleasurable, not afflictive.

208-What is another explanation about the suffering of the soul by fire?
Another explanation is that the soul, upon seeing the fire and perceiving it as harmful, becomes distressed. Saint Gregory mentions that the soul suffers because it sees itself burning.

209-Is fire truly harmful to the soul?
Saint Thomas maintains that if fire were not truly harmful, the soul would be mistaken in perceiving it as such, which does not seem reasonable, especially in the case of demons, who have great intellectual acuity.

210-What conclusion does Saint Thomas draw about the nature of fire?

He concludes that corporeal fire is indeed harmful to the soul, as this fire, by divine power, acts as an instrument of divine justice.

211-How can corporeal fire affect an incorporeal substance like the soul?
Saint Thomas explains that the soul does not suffer direct alteration or destruction from fire; rather, it suffers because the fire prevents its natural inclination of not being subject to a specific place.

212-What kind of suffering does the soul experience due to this limitation?
The soul experiences an interior sadness, as it perceives the fire as contrary to its nature, which causes distress.

213-What is the greatest affliction of the damned souls, according to Saint Thomas?
The greatest affliction is their separation from God.

ENDNOTES

[1] Cfr. DE AQUINO, TOMÁS. *Cuestiones disputadas sobre el alma. Traducción y notas de Ezequiel Téllez Estudio preliminar y revisión de Juan Cruz Cruz*. Ediciones Universidad de Navarra, S.A. (EUNSA). Edición digital de @elteologo Agosto de 2014. Pages LV-LXXII.

[2] The agent intellect, according to Thomistic explanation, does not itself elaborate concepts and ideas; its role is rather to make the potential ideas in our mind become intelligible in act, that is, to make them truly understandable. It does this by abstracting forms or intelligible species from the images or phantasms, which are the sensible representations of the things we perceive.

To understand when the agent intellect performs this function, it is helpful to think of the process in two parts:

1. Reception of the phantasms (sensible images): When we perceive something through the senses, our possible intellect cannot directly comprehend these images or phantasms because they are bound to their individual and material characteristics. For example, when seeing a flower, the mental image we generate is filled with individual details (its color, specific size, etc.).

2. Abstraction of the universal concept or idea: This is where the agent intellect comes in. Its function is to "illuminate" or abstract the universal form (for example, "flower-ness" in general) from the particular case. In doing so, it removes the material and contingent aspects of the sensible image, leaving only the universal essence of "flower." This abstract essence is then captured by the possible intellect, which understands it as a concept or idea.

The possible intellect is the one that formulates and holds the universal concept of "flower."

The agent intellect does not formulate the concept as such, but performs the work of abstraction by removing the particularities of the sensory images (such as the specific features of a particular flower) and extracting the common form, that is, what is essential and universal in all flowers. This process of abstraction is what makes it possible for the possible intellect to later grasp and formulate the concept of "flower" as something universal.

Therefore, the agent intellect prepares the ground for the possible intellect to formulate the universal concept of "flower." Both are necessary for the process, but the possible intellect is the one that, in the end, elaborates and formulates the abstract concept, as in this case the concept of "flower."

[3] At first glance, it may seem strange to compare things as different as bodies and souls. However, in the philosophical and theological context in which this argument is presented, the connection between the two is understood through the concept of perfection in terms of their purpose and order within the universe.

To explain this further: in scholastic philosophy, especially in the thought of Saint Thomas Aquinas, each being possesses a perfection or fullness that fulfills it in its own mode of existence. Some beings are considered to have a greater perfection when they fulfill a function that is more noble within the order of the universe. Thus, although bodies (such as celestial bodies) and souls (such as the human rational soul) are distinct in their nature, they can be compared in terms of the nobility of the perfection they achieve.

Here, celestial bodies are seen as particularly noble and perfect because, in the medieval view, they do not decompose, do not suffer corruption, and move eternally in regular orbits, making them more "perfect" or "complete" in their physical order than earthly bodies. However, the human rational soul is considered superior in another sense: its perfection lies in its capacity for intellectual knowledge and love, in its ability to know eternal truths, which is understood as an even greater form of "perfection."

The connection between the two, then, is found in the idea of "perfection" according to their purpose and function in the universe. This comparison seeks to understand whether the human body, perfected by a rational soul, holds a different or even greater kind of perfection than a celestial body perfected by a spiritual substance that moves it. In this logic, the human soul, though united to a corruptible body, performs a higher function due to its intellectual capacity, which is considered the highest purpose in the created order.

Therefore, the argument does not compare what body and soul are, but rather the type and degree of perfection each one reaches according to the role it fulfills in the order of the universe.

[4] In this context, *contrariety* refers to the presence of opposing or conflicting qualities within the human body, which makes it susceptible to change, decay, and ultimately destruction. In Aristotelian-Thomistic philosophy, sublunary bodies (that is, those that exist in the terrestrial world) are subject to contrary qualities, such as heat and cold, or moisture and dryness. These opposing qualities interact, and in doing so, they cause

the wear and corruption of physical bodies.

[5] Work falsely attributed to Saint Augustine but, in fact, belonging to Alcher of Clairvaux, a Cistercian monk.

[6] The text suggests that not all parts of the body are organic because the soul, in the Aristotelian conception, acts as the vital principle that animates and gives form to a body that is structured for life. The parts of the body that are considered "organic" are those that have a vital function and are interrelated in a way that contributes to the functioning of the organism as a whole. For example, organs such as the heart, lungs, or liver have specific roles that enable the life of the human being.

However, there are parts of the body, such as hair, nails, or even some bone structures, that are not directly related to life or organic function in the same sense. These parts are not "organic" in the sense that they do not actively contribute to the vital functions of the organism. Therefore, the soul, as the principle of life and substantial form, cannot be present in these parts in the same way that it is in the organs that effectively sustain and enable life.

Thus, the idea is that the soul is present in the parts of the body that have a vital function and are capable of participating in the process of life, while in the non-organic parts, the relationship with the soul is different, as they do not possess that capacity for animation and vitality that defines the organic.

[7] The quantitative division refers to the way we conceive of a whole in terms of its physical dimensions or quantities. This perspective views the body as an entity that can be divided into parts according to its size, volume, or extent. In this sense, something can be considered "total" in relation to its size, where the totality is understood as the sum of its parts. For example, an object like a table is a whole by virtue of its size and the parts that make it up (the tabletop, the legs, etc.). However, this conception of totality is more superficial, as it does not address the essence of the object itself, but only its physical arrangement. In the case of the soul, this division does not apply in the same way, since the soul cannot be measured or divided quantitatively; its nature transcends physical dimensions.

The essential division focuses on the intrinsic relationship between form and matter that constitute a compound. In this context, a being is considered a whole by virtue of its essence, where the form gives identity and specificity to an entity, and the matter is the substrate in which this form is actualized. For example, in the case of a human being, the soul is the form that gives life and specificity to the body, thus constituting a single entity. This perspective emphasizes that for a compound to be a whole, there must be a meaningful integration of its parts in relation to its essence. This means that the parts not only exist together, but are constitutively united by the form that grants them unity and meaning. In this sense, the soul is present in every part of the body as its form, which makes each part participate in the identity of the whole.

The division by power or virtue refers to how a form can act or be realized through its parts. This conception of totality focuses on the capabilities and operations that an entity can exercise, considering that different parts may have different roles and functions in relation to the activity of the whole. For example, in an organism, the heart and lungs have specific functions that contribute to the overall health and functioning of the body. Thus, the soul, although present in every part of the body, does not exercise its power uniformly in all of them. Some parts are responsible for functions that require a more intense manifestation of the soul's virtue, such as understanding or will, while other parts perform more basic and mechanical functions. This dimension highlights that totality cannot be understood solely from the perspective of form or matter, but must also consider the various capabilities and roles of the parts in the functioning of the complete entity.

Together, these three modes offer a rich and multifaceted understanding of totality, helping to unravel the complex relationship between the soul and the body, as well as the nature of beings in general. By considering these modes, a more complete view is achieved that not only limits itself to the physical structure but also takes into account the essence and operative capacities of entities.

[8]The phrase "each genus has a single principal contrariety" refers to the idea that within each genus or category of qualities, there is a fundamental pair of opposites that represents that category. In philosophy, a "genus" is a broad category encompassing different species or types, and a "contrariety" is a fundamental opposition between two terms.

In this context, the text suggests that if sensory powers (such as touch or sight) are diversified according to different genera of qualities (for example, colors in sight or temperatures in touch), then each genus would have a principal contrariety. This would imply that each sense should have different powers to perceive these contrarieties. However, the text points out that this is not the case in all senses; for instance, in touch, there are no clear divisions between opposites such as "hot" and "cold," "soft" and "hard," all of which are perceived without the need for distinct powers.

[9]In this context, "contrarieties" refers to the presence of opposing or conflicting elements that, in other beings, may lead to corruption or

dissolution. Below are some examples of what could be interpreted as "contrarieties" within the soul:

1- Will vs. Appetites: Appetites or desires can sometimes seem opposed to rational will, as a person might desire something that reason deems harmful. Although this appears to be an internal contrariety, the argument suggests that these desires and decisions do not imply a conflict that would corrupt the essence of the soul.

2- Intellectual Knowledge vs. Sensory Knowledge: Sensory perception and intellectual knowledge may lead to different judgments. For instance, the intellect might recognize an object as harmful, while the senses find it pleasurable. However, this difference in judgments does not corrupt the soul but remains without causing division in its essential being.

3- Love vs. Hatred: In emotions, one might feel love for one aspect of a situation and hatred for another. This duality might seem like a contrariety, but the human soul can hold both dispositions without implying corruption or a division in its essence.

4- Reason vs. Impulsive Emotions: Reason often seeks to control or moderate impulsive emotions, such as anger or sadness. Though this relationship may appear to be a contrariety, it does not manifest as a destructive division in the soul that would lead to its corruption.

The argument holds that these apparent contrarieties do not compromise the essential unity of the soul. Unlike material realities, which can decompose due to conflicts between opposing elements, the soul retains a unified nature and, therefore, remains incorruptible.

[10]Saint Augustine retracted several statements and concepts he had previously expressed in his writings in his work *Retractationes*, especially concerning the nature of Hell. One of the key points mentioned is the notion of Hell as a physical place, in the sense of a geographical space beneath the earth.

Specifically, in his commentary on Genesis *(De Genesi ad litteram)*, Augustine acknowledges that his understanding of Hell's location as a concrete place might not be entirely accurate and that it is more appropriate to consider it in terms of a state or condition of separation from God.

Additionally, Augustine also notes that the depiction of Hell in terms of physical suffering and a place of torment should be understood more symbolically than literally. In this sense, his retraction suggests that Hell should not be viewed solely as a specific geographical space but as a spiritual reality involving the absence of divine grace and the suffering resulting from that separation.

This reconsideration influences the theological interpretation of Hell's nature and how arguments about it should be approached, as Thomas Aquinas points out in his response to arguments that seem to assert that the sensitive powers remain in the separated soul.

[11]"Individual intentions" and "universal intentions" are key terms in the theory of knowledge.

1- Individual intentions. These are the specific forms or mental representations that the soul acquires through the senses when interacting with particular, singular objects in the world. These intentions are individual because they are related to direct and particular sensory experiences, such as a specific person, a particular object, or a unique event. These intentions reside in the soul's sensitive faculties (memory, imagination) and depend on the presence of the body and its sensory organs. When the soul is separated from the body, it loses the capacity to sustain these individual intentions, as it no longer has direct access to the sensitive faculties.

2- Universal intentions. These are mental representations of the common or essential nature of objects, abstracted from their particular characteristics. When the intellect knows something, it abstracts a universal "species," an idea that does not refer to any particular individual but to the general nature of a thing (for example, "humanity" rather than a specific person). Universal intentions reside in the understanding and, unlike individual intentions, can subsist in the soul when it is separated from the body. However, these intentions do not allow for knowledge of the singular, as they are abstract and general.

The text explains that the separated soul cannot know the singular through the forms it acquired in the body or through forms that might be divinely infused, as these forms or universal species relate to objects in general, not to particular cases. This reveals the limitation of singular knowledge for the soul in its state separated from the body.

[12]The quintessence, in ancient philosophy and cosmology, represented a special element from which celestial bodies, considered incorruptible, were formed. This element was often known as ether, a light substance that filled space and through which it was believed that light and other cosmic forces were transmitted. In contrast to the four traditional elements—fire, air, earth, and water—that composed earthly bodies and were seen as corruptible, the quintessence belonged exclusively to the supralunar world. This realm was conceived as one of perfection and eternity, in opposition to the sublunar world, where corruption and change were inevitable. Thus, the quintessence stands as an ethereal and divine element, reflecting a

hierarchy in nature where it occupies a superior position, associated with the eternal and the transcendent.

www.ingramcontent.com/pod-product-compliance
Lightning Source LLC
Chambersburg PA
CBHW082244220526
45469CB00009B/2872